EXPERIMENTAL PHYSIOLOGY

CONTRIBUTORS

C. G. Ingram, M.B., CH.B.
Lecturer in Physiology,
St. Salvator's College, University of St. Andrews

J. H. Taylor, M.B., CH.B, D.P.H., M.R.C.P.E.
Senior Lecturer in Physiology,
St. Salvator's College, University of St. Andrews

EXPERIMENTAL PHYSIOLOGY

edited by

B. L. ANDREW D.SC.

Senior Lecturer in Physiology,
University of Dundee

EIGHTH EDITION

E & S LIVINGSTONE LTD
EDINBURGH AND LONDON
1969

SBN 443 00618 0

Seventh Edition - - - - - - 1965
Eighth Edition - - - - - - 1969

Made and Printed in Great Britain

PREFACE TO THE EIGHTH EDITION

WHEN first published in 1936, this book was designed by its author, Professor G. H. Bell, to serve as a laboratory manual for medical students taking their course of experimental physiology at Glasgow University. The editions that followed kept pace with the gradually enlarging curriculum and by the sixth edition in 1959 it contained rather more experiments than would normally be included in a course for medical students and so could be used either as a laboratory manual or as a departmental source-book in the planning of a variety of courses. This process has continued in the seventh and the present edition. Used selectively, the book serves as a laboratory manual for medical, science and dental students at the Universities of Dundee and St. Andrews. Every department uses the special skills of individual members of its staff to provide technically difficult demonstrations which can greatly enliven a course. As apparatus improves it has been possible to transform some demonstrations into class experiments.

In the present edition the chapter on muscle and nerve has been expanded to include some important neurophysiological experiments; some of these are suitable for science students in their second year. The problem of recording rapidly occurring events without the use of photography can be eased, though rather expensively, by the introduction of the storage oscilloscope. Methods used to study muscle and nerve can be selected either from direct mechanical recorders or from transducers and oscillographs. Technical difficulties in respiratory physiology can be reduced by the use of respiratory gas analysers using physical rather than chemical methods and many have welcomed the eviction of the mercury manometer by the pressure transducer in cardiovascular experiments. A new chapter on alimentary tract and kidney fills a gap which had become rather obvious. New experiments and figures have been added to most of the chapters and I thank reviewers and colleagues for advice on these matters.

It is a pleasure to thank Professor G. H. Bell for his continuing interest in 'Experimental Physiology'. Thanks also to Dr. Taylor and Dr. Ingram, who have charge of courses at St. Andrews University, for contributions, and to Mr. G. C. Leslie, Dr. J. M. Patrick, Mr. J. Thompson and Mrs. M. Glenday from the department here in Dundee for help in the production of this edition.

B. L. A.

Department of Physiology
The University, Dundee
April, 1969

CONTENTS

CHAPTER ONE

INTRODUCTION TO APPARATUS AND INSTRUMENTS

During his course of experimental physiology, the student is faced with a wide variety of apparatus and instruments, some drawn from physical and chemical laboratories which he may have seen or used before, others specifically developed for physiology. He will be expected to become skilled in the use of some of the apparatus. How much he uses himself and how much is demonstrated to him depends on the person who organizes the course; his decisions will be influenced by the number of demonstrators and the amount of technical support available to keep the apparatus in working order.

The student should think about the fitness of the instruments for the purposes they serve in an experiment and try to assess the extent to which his record is a faithful representation of the activity of the tissue or biological system that he is examining.

Instrumental complexity varies from the easily understood lever attached by a thread to the ventricle of an isolated heart to the 'black box' which writes out the electrocardiogram; the intermediate steps in the operation of the latter device will probably be as obscure as the intermediate optical steps in the operation of a microscope are to the average histologist.

Nevertheless, to reach a position in which he can cease to be distracted by the apparatus and able to concentrate on the biological phenomena, the student must gain a certain mastery over his apparatus; enough, at least, to know if it is functioning correctly and able to record the biological quantities involved.

In this chapter a short description is given of the common recording and measuring instruments used in the experiments described later in the book. In this way duplication of instrumental descriptions in later chapters is avoided. A subscript number after an instrument refers to the list of manufacturers in appendix A; the bibliography in appendix B is arranged in groups of books to support individual chapters. The book is arranged so that the student can insert his records and comments close to the paragraphs describing the experimental procedures. It is usually not necessary or even desirable to write up detailed accounts of the procedures, whereas results, comments and conclusions are an essential part of the training in scientific work. Observations should be written out in a notebook, not on odd pieces of paper, and should be tabulated and condensed as necessary before insertion into this book.

Surgical Instruments

The student should keep a set of dissecting instruments for his own personal use; the list given below is the minimum requirement. They should be made of stainless steel and sharpened as necessary.

1 5 in. straight stitch scissors with sharp points
1 small dissecting scissors with sharp points
1 large blunt-nosed forceps
1 dissecting forceps, fine points
1 mounted needle
1 seeker
1 curved triangular needle
1 large scissors for rough work

Usually the laboratory will provide other instruments, such as bone clippers and scalpels, on loan. To get long life out of the smaller instruments it is essential that they are only used for delicate work. For nerve dissection work, watchmakers' forceps$_3$ with needle points are necessary.

The Electrical Stimulation of Muscle and Nerve

A muscle cell will contract, or a nerve fibre will initiate an impulse, if an electric current of suitable size and duration is passed through the tissue. The conductors which are used to bring the current to the tissue are called *Electrodes* and in many experiments are silver or platinum wires applied directly to the tissue. The apparatus which produces the electric currents is called the *Stimulator*.

1.1 The Stimulator

Three types of stimulator will be used, a simple square wave generator called 'Student Stimulator' (Palmer type H47), a Double Pulse Stimulator (Electrophysiological Instruments$_{27}$ type 2CROS), and an isolated stimulator unit (Devices$_{11}$ Isolated Stimulator Mk. IV).

The stimulating current from a square wave stimulator is turned on and off abruptly and can be represented graphically as in Fig. 1.1. The line indicates the voltage between

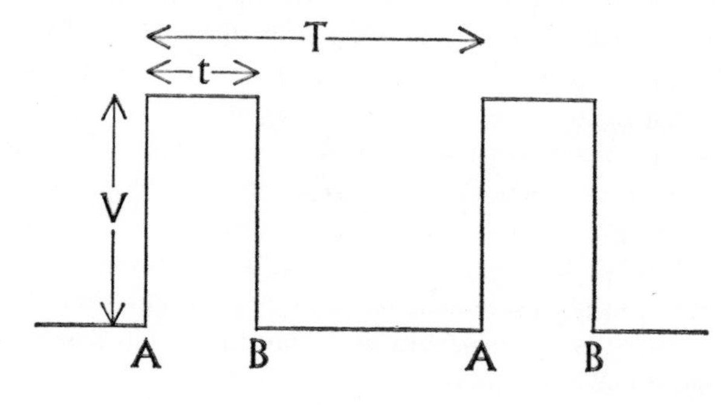

FIG. 1.1

the output terminals of the stimulator; the current that will flow through the tissue will depend on the electrical resistance of the tissue (and on the internal resistance of the stimulator). The diagram indicates that the voltage between the output terminals suddenly changes from zero to V at point A, stays constant for interval t and then returns to zero at point B. V is called the pulse height or stimulus strength, and t is the pulse length (width) and is measured in milliseconds; if the pulse is one of a regular series then it has a repetition rate of $\frac{1}{T}$ pulses per second.

1.2 The Student Stimulator

Pulse strength (V). The central upper control is graduated 0–10 and this corresponds to an output voltage range of 0–10 volts. Pulse length (t). The upper right hand switch has two positions, the long pulse is 1·5 milliseconds and the short 0·5 milliseconds. Pulse Rate (1/T). The lower right hand control is a 12-position switch which covers repetition rates between 1 pulse in 20 seconds (·05) and 100 pulses per second (100).

There are two modes of operation for a stimulator, either it produces a single shock when commanded to do so by the experimenter, this command circuit is called the Trigger Circuit, or it produces a long train of pulses at a steady rate. The lower left switch has three positions and is called the selector switch because it selects in which mode the stimulator operates. In the left hand position marked 'EXT. TRIG.' the stimulator will produce one pulse each time the left hand terminals (EXT. TRIG.) are connected together momentarily. The generation of this pulse is indicated by a flash on the central lower neon indicator. The voltage of this pulse is decided by the setting of the strength control and the length by the long/short switch. The pulse rate control is inoperative in the EXT. TRIG. position of the selector switch.

When the selector switch is moved to the right hand 'REPEAT PULSE' position the stimulator operates in the second mode and produces a train of pulses at the rate indicated by the setting of the Pulse Rate control.

When the selector switch is in the central position, PULSE OFF, the output terminals are disconnected from the stimulator, so that even though the neon tube flashes no output is produced at the terminals.

Operating Instructions

1. Set selector switch to PULSE OFF position.
2. Insert mains plug into bench outlet, switch on mains and also upper left switch (on/off) on stimulator.
3. Set Pulse Rate switch to 5, wait until the neon light flashes at this rate; this delay is the warm-up time of the valve and is about 10 seconds.
4. Set 'Strength' control to zero and connect the electrodes to the output terminals.

The next move depends on the type of experiment, whether you want single pulse operation or not. If you wish a pulse to be initiated at a point during the rotation of the kymograph drum, connect EXT. TRIG. terminals to the drum contact-maker terminals.

Each time the contact-maker arm of the kymograph strikes the contact-maker switch, it will short-circuit the EXT. TRIG. terminals and initiate a pulse; for this mode the selector switch should be in the left hand position. If you wish to initiate pulses manually at your own command, connect a hand switch or push-button switch across the EXT. TRIG. terminals and press when a pulse is required. This trigger switch must have a high insulation resistance and low capacitance in the 'off' position and low contact resistance in the 'on' position or it will fail to trigger. If your experiment requires a train of shocks, as for example when stimulating the crescent of the frog's heart, leave the selector switch in the central 'PULSE OFF' position and set the pulse rate switch at the desired position. The train of pulses can now be started and stopped by moving the selector position from PULSE OFF to REPEAT PULSE and back. Switch off the instrument by the top left hand switch at the end of the experiment.

1.3 The Double Pulse Stimulator

This stimulator ($2CROS_{27}$) during a single operating cycle emits a triggering pulse to start the time base of the oscilloscope and thereafter two stimulating pulses. The pulse height, pulse width and delay after the triggering pulse of each of the two pulses can be adjusted at will. The cycle of operation may repeat itself at a frequency between one cycle in five seconds and one hundred per second or it may be triggered from an external contact to occur once. When in the free running mode it can be started and stopped by an external contact. The block diagram of the stimulator is shown in Figure 1.2 together with connections to an oscilloscope. To gain familiarity with the controls a preliminary exercise with an oscilloscope is described in paragraph 1.10.

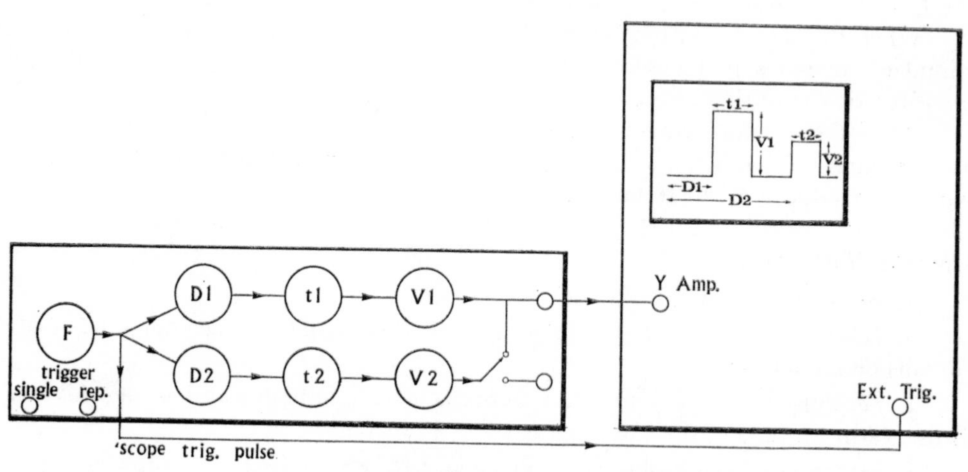

FIG 1.2

A double-pulse stimulator is shown connected to an oscilloscope as described in para. 1.10. For single-cycle operation a jack connected to a push-button switch (press to make contact) is inserted into the jack socket marked 'single'. If the jack is inserted into the 'rep' socket, the stimulator stops but runs repetitively so long as the press-button is depressed.

1.4 The Isolated Stimulator Unit

This stimulator$_{11}$ can be triggered either by the push button on the panel or by a 12 volt square wave. It emits a single square wave whose length can be varied between 0·05 msec and 1 sec, and whose voltage can be varied up to 90 v. It is used in experiments where the range of pulse width and voltage obtainable from the 2CROS stimulator is insufficient. To obtain a train of pulses it is necessary to trigger it repetitively from either a student stimulator or a 2CROS.

1.5 Stimulus Isolation

In experiments in which a square wave stimulator is used in conjunction with a high gain amplifier it is important that the stimulus voltage should appear between the stimulating electrodes and not between them and earth potential. A battery operated stimulator such as the double pulse stimulator 2CROS$_{27}$ or the Devices Isolated Unit Mark IV is already isolated from earth so no difficulty will arise, but a mains driven stimulator such as H47 has one output terminal earthed and therefore requires an isolator unit between output and electrodes. If isolation is inadequate, part of the stimulus voltage is injected into the input of the amplifier of the pick up electrodes and produces a large shock artifact and circuit paralysis. A radiofrequency coupling unit is described by Donaldson (1958).

An isolated circuit, because of the very high impedance between it and earth, particularly if the electrodes are applied to a delicate nerve filament, is liable to pick up interference from the mains and voltages sufficient to stimulate the tissue at mains frequency can be introduced by the field of a nearby oscilloscope or other electrical apparatus. Electrical shielding of these circuits is therefore important; it may be necessary to include the isolated stimulator within the shield surrounding the preparation.

1.6 Recording Instruments

These instruments vary both in their sensitivity and in their ability to follow slowly or rapidly occurring events; the choice of instrument depends on the tissue or system under study. The greatest difficulties are found in experiments on the nervous system where steady and rapidly-changing potentials of small size may have to be recorded simultaneously. This requires an amplifier and recording system with a level frequency response from zero frequency (d.c.) to 10 kilocycles and a maximum sensitivity sufficient to show deflections of a few microvolts. The instrument of choice for this task is the cathode-ray oscilloscope.

For events occurring more slowly, such as the potential changes of the electrocardiogram or the electroencephalogram, the ink-writing oscillograph is preferable$_{11}$; such instruments commonly have a frequency response up to 100 cycles per second and provide a permanent record immediately.

Where electrical amplification is not needed and a direct mechanical link between the writing point and the driving system is convenient, as in recording respiratory movements from a spirometer or stethograph, ink-writers on glazed paper are used.

In intermediate situations where there is only just enough power available to work the recording lever, as in the frog heart and muscles, it is traditional to use a stylus writing on a smoked paper surface. The student will be introduced to a variety of such mechanical lever systems; the choice of the lever is influenced by such considerations as the magnification required and the amount of power which can be taken from the system without distorting its action. Some levers introduce gross distortions, e.g. the frog muscle lever may oscillate when it hits the limit stop after rapid relaxation of the muscle. If the recording device puts such a load on the tissue under observation that its response is modified it is better to use a device which will convert the change in the tissue into an electrical variation which can then be amplified and used to drive an ink-writing oscillograph or a cathode-ray oscillograph. These devices are called *transducers* and are commonly used to convert such quantities as blood pressure, air flow, heart sounds and muscle movements into electrical potential changes.

1.7 Transducers

These devices convert some physical measurement such as temperature, force, displacement, pressure or acceleration into an electrical signal which can be used to drive a recording instrument. Usually the electrical output is a linear electrical equivalent of the quantity measured over a limited range (which is specified for the particular transducer).

Transducers are used in this book to measure force, pressure, displacement and temperature and short notes are given below on their principles of operation.

Force. The force to be measured is applied to a stiff mechanical element which moves only slightly over the operating range and is protected from overload by a mechanical stop. The distortion of the device is transmitted to silicon strain gauge elements which change their electrical resistance. The resistance change is converted by the associated bridge circuit into a voltage output and this is used to drive the recording device or oscilloscope. Usually two silicon elements are used, forming a half bridge, the applied strain increases the value of one element and decreases the value of the other.

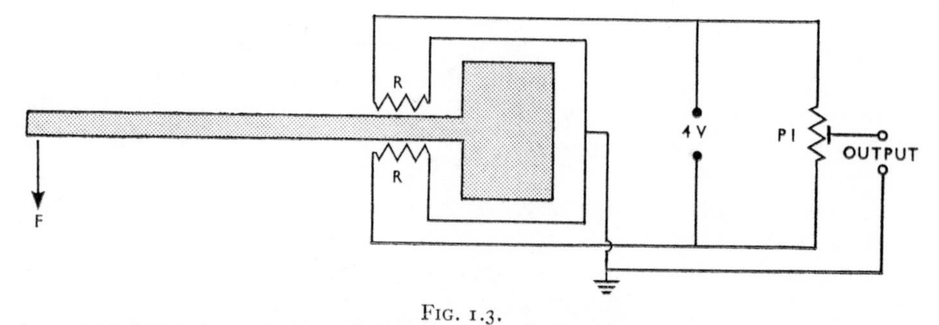

FIG. 1.3.
A cantilever beam, shown shaded, has two strain-sensitive resistances (R) cemented to it. A force (F) when applied to the beam, bends it slightly, changes the values of R, and disturbs the balance of the bridge circuit formed with potentiometer P1, which has been previously adjusted to give no output when F is zero.

The circuit is arranged as in Figure 1.3. The 4v bridge energizing voltage can be supplied by a dry battery and this should be housed inside a metal screening box with balancing potentiometer P1. The Devices[11] 2STO2 transducer operates over a force range of 1 kg. wt. and when energized with 5v gives an output of 0·4 mV per g. In this transducer the applied force bends a cantilever beam. The Ether[31] model operates over 100 g. wt. and gives an output of $1\frac{1}{2}$ mV per g.

In some force transducers wire strain gauges are used instead of silicon strain-sensitive elements but the output voltage per g. force is substantially less.

Pressure. To measure pressures at points in the cardiovascular system[11,31], a column of liquid in a rigid-walled catheter is led to a small chamber, one side of which is a stiff metal membrane. Attached to the membrane are silicon or wire strain gauges in either half or complete bridge configuration. The bridge is energized by a few volts and the signal after amplification is fed into the recorder. For the lower pressures which have to be measured in some respiratory experiments, pressure transducers operating on photoelectric principles[27,28], are more convenient, because of their greater sensitivity.

Displacement. Two methods are commonly used. For very delicate preparations, such as the frog's heart, a photoelectric transducer will give a large electrical output. These transducers, which do not appear to be available commercially, can be made quite easily. The heart movements are transferred to an opaque shutter which lies between a low voltage lamp and a silicon phototransistor. The circuit is shown in Figure 1.4. The metal lever of a Starling heart lever can be used as the shutter, in which case it is only necessary to fit a lamp assembly on one side of the lever and a phototransistor on the other. Room lighting must, of course, be excluded from the phototransistor. Where more power is available to drive the transducer, for example when measuring the displacement of the bell of a spirometer, linear or rotary potentiometers are suitable[31].

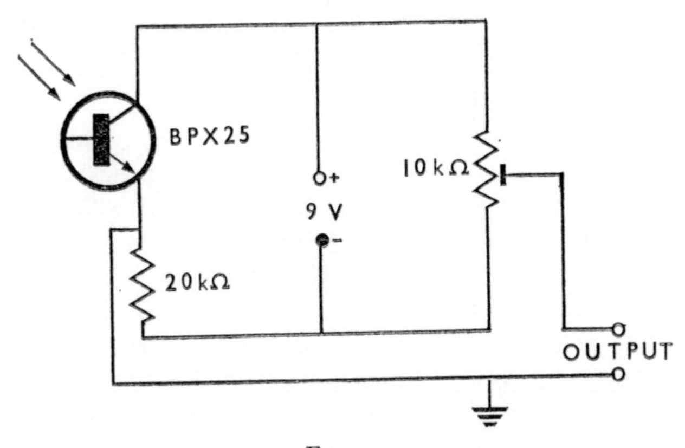

FIG. 1.4

Temperature. The simplest device is the thermistor[6,19], which changes electrical resistance with temperature, see circuit Figure 1.5. The thermoelectric method, which until recently required an expensive and easily damaged galvanometer is now simplified by the introduction of transistor amplifiers which enable a robust meter to be used.

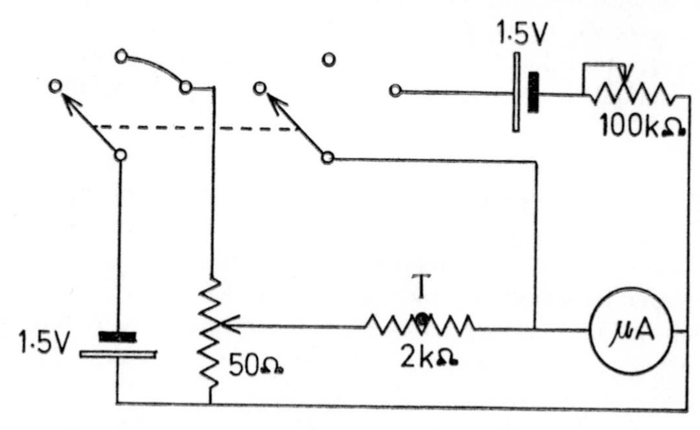

FIG. 1.5

The thermistor, T, has a resistance at 25° C (R_{25}) of 2kΩ. Setting-up procedure. The switch is shown in the left-hand, 'OFF', position. Place tip of thermistor in a beaker of water at the lowest temperature to be recorded experimentally; in the case of expt. 6.15, use an ice/water mixture. Turn switch to mid-position and adjust the 50 ohms potentiometer until the 100 μA meter reads 75 μA. Turn to right-hand, 'ON', position of the switch, and adjust the 100k. ohms variable resistance until the microammeter reads zero. Now put thermistor tip in water at several temperatures, together with a thermometer, and construct a calibration curve relating μA to °C. When used to measure skin temperatures a narrower range of temperature is encountered and the scale can be expanded; adjust the 50 ohms potentiometer to give full-scale deflexion with the switch in the mid-way position and the thermistor tip in a beaker of water at room temperature. A calibration curve is then constructed as before.

1.8 The Cathode-Ray Oscillograph

The cathode-ray tube enables the visualization and photography of electrical potentials and it can respond so quickly that there is no practical limit to its application in physiology. Furthermore, when used in conjunction with amplifiers it offers so little loading that the activities of single cells can be recorded. The student is not expected to know the details of the circuits operating the cathode-ray tube, but it is necessary to describe certain features of the input and output circuits so that it may be used correctly.

The cathode-ray tube contains an electron beam generator, called the gun, whose output passes through zones where it may be deflected by potentials applied to plates. Usually there are two sets of plates orientated at right angles to one another. The deflexions in the horizontal plane are called X deflexions and in the vertical plane Y deflexions. The electron beam, after passing through the deflexion zones, travels to the flat far end of the tube, which has a surface coated with a material which emits visible light when the electron beam impinges on it. The properties of this material, or phosphor, are important in terms of colour (fluorescence colour) and in the length of time the phosphor will glow after the electron beam has ceased to activate it (phosphorescence colour and persistence time). If an oscilloscope is to be used for a wide variety of experiments it is convenient to have two or three tubes, each with a different phosphor. Photographic recording is easiest with a blue phosphor with short persistence (1 to 5 msec) and is essential if a moving film camera is used: in this situation perceptible afterglow will produce smudging of the photographic record. For general visual observation a green or yellowish green phosphor is used; the green may have a persistence of 100 msec and the yellowish green 2 to 10 seconds. For slowly occurring non-recurrent phenomena a longer afterglow is needed; these tubes have a bluish white fluorescence which produces a yellowish green phosphorescence persisting many seconds. Methods developed to permit the simultaneous observation of two cathode-ray beams include multigun tubes in which two or more quite independent guns and ray deflector systems are aimed at a single phosphor surface, and split-beam tubes in which the beam from a single gun is divided and guided through independent deflector systems.

The variable persistence oscilloscope[4] has a cathode-ray tube whose afterglow properties can be varied at will so that the persistence of the image can be matched to the time of sweep of the time base. The storage oscilloscope[5] maintains the afterglow for up to an hour, and it is possible to display a stored image on one half of the tube face while a time base with normal afterglow is used on the other half of the tube face. The variable persistence and the storage oscilloscopes have wide applications in experimental physiology and greatly reduce the need for photographic recording; only their relatively high cost prevents their widespread use.

The oscilloscope controls can be grouped into (a) gun controls (b) beam deflexion controls (c) amplifier controls and (d) time-base controls. Introductory experience in the use of the instrument should be gained on a simple oscilloscope with only a few facilities; the complex double-beam oscilloscopes have 40 or 50 controls and are baffling to the beginner.

(*a*) *Gun Controls.* These are Brightness, Focus and Astigmatism, and permit the size and brightness of the luminous spot on the phosphor to be adjusted. The spot must not be allowed to remain stationary at high brightness or the phosphor may be damaged.

(*b*) *Deflexion Controls.* These are X and Y shifts.

(*c*) *Amplifier Controls.* To match the sensitivity of the cathode ray-tube to the electrical size of the input signal, amplifiers with variable gain and sometimes variable frequency response are provided for the Y deflexion plates. The amplifier control is calibrated in voltage input required to produce 1 cm deflexion of the beam and the range provided in an inexpensive oscilloscope is from 50 volts per cm to 10 millivolts per cm. An oscilloscope with an input sensitivity of 1 mV/cm or 100 μV/cm costs substantially more. Sensitivities greater than 10 mV/cm are required for several of the experiments in this book, in these instances the extra sensitivity can be obtained by using a preamplifier[34,35] between the pick-up electrodes and the input to the oscilloscope amplifier. This arrangement has the advantage that a battery driven preamplifier can be mounted close to the preparation, while the oscilloscope, which may generate strong magnetic fields at mains frequency, may be set up at a distance where it will cause less interference.

(*d*) *Time-base Controls.* It is usual to display the input signal in terms of time, so that a graph is produced with time as the horizontal or X axis. To produce steadily moving deflexions of the spot along the X axis a Time-Base Unit is provided. It has a range switch marked in units of time taken for the spot to move one cm. This range usually extends from seconds per cm to μsec per cm.

The time-base will run in two modes, spontaneously (or free-running) or triggered by an external signal; this external signal may be some feature of the signal entering the Y amplifier or it may be a synchronizing signal entering the time-base via the 'External Trigger' terminal. Switches are provided to determine the source and polarity of the trigger pulse. Whether the time-base runs spontaneously or only when triggered depends on the setting of the Stability Control. The voltage value of the trigger signal which sets the time-base moving is adjusted by the Trigger Level control. Sometimes it is advantageous to apply one signal in the usual way to the Y deflector plates while a related signal is applied to the X deflector plates, this arrangement is called XY recording. The time-base circuit is switched off and the X axis signal connected to the X plate amplifier input terminal.

Choice of Oscilloscope

Many experiments described in this book do not call for a very complicated oscilloscope. A double-beam model, such as the Telequipment D52[4], supplemented with a pre-amplifier[34] where a sensitivity greater than 10 mV/cm is required, is satisfactory.

1.9 Photography of Cathode-Ray Tube Traces

Single time-base sweeps can be photographed with only a slight reduction in size by the 'Polaroid' process, and cameras are available with a sliding back which allows several traces to be positioned one above another on a single print. If the events being recorded cannot conveniently be positioned on a single time-base sweep then the Moving Film method is employed. For this method, the time-base is switched off and the signal applied via the Y amplifier so that the spot is deflected in the Y axis only. The image of the spot is photographed on film or sensitized paper which moves at a steady speed through the camera in a plane at right angles to the plane of the Y deflexion of the tube. In effect the film drive motor provides the X deflexion, and a continuous record with any convenient time-base from millimetres to metres per second can be obtained. This method is usually employed to record activity in peripheral nerve fibres.

1.10 To Display the Output of the Double Pulse Stimulator on the Oscilloscope

Turn on the oscilloscope, set time-base speed at 2 msec/cm with external triggering. Turn Trig. Level fully anti-clockwise, but short of the 'auto' position. Set 'Trig Selector' at External—negative. Adjust time-base stability control so that the time-base just does not run spontaneously. Connect black 'Scope Trig. Pulse' socket on the stimulator to the oscilloscope earth and the white 'Scope Trig. Pulse' to the external trigger terminal of the oscilloscope. Set Frequency Control of stimulator at 10 per second and turn on the stimulator.

Adjust the oscilloscope Trigger Level control until the time-base sweeps. If in any doubt turn off the stimulator, the time-base should then stop. If it does not stop reset the stability control until it just stops, turn on the stimulator, adjust Trig. Level until it restarts. During the warming-up period, i.e. the first ten minutes of operation, the operation of some time-base and gun-control circuits changes slightly. Connect the stimulator output (C1 + C2) to the Y1 amplifier input. Set Y1 sensitivity at 1 volt per cm. Adjust stimulator controls as follows:

Delay C1 $>$ 1 msec, C2 $>$ 1 msec,
Width C1 minimum
C2 minimum
Intensity range C1 0–3v dial 10
C2 0–3v dial 8

You should see two square wave pulses on the time-base (see Fig. 1.2). Explore the effects of turning the delay, width and intensity controls of each channel.

1.11 The Kymograph

The kymograph[1,33] consists of an electrically-driven gear box with a vertical spindle carrying a drum. The gear box has a clutch and controls which determine the final speed of the spindle. The clutch *must* be disengaged before changing gear.

The choice of drum speed is important and this is usually specified in terms of peripheral paper surface speed. The kymograph may have up to 12 gear positions and a separate control for fine speed adjustment to give a range of rotation time between one second and thirteen hours; with a 6 in (15 cm) diameter drum this corresponds to a paper speed between 500 and 0.01 mm per second. As several models of kymograph may be available in the same laboratory, the student should explore the controls of his kymograph at the beginning of the session.

The vertical spindle carries two horizontal arms which can move stiffly on the spindle. When the drum rotates these arms briefly close a contact-maker mounted on the base. This device, when connected to the external trigger terminals of a stimulator can initiate a stimulus at a particular point of the drum rotation.

By convention, the drum rotates in a clock-wise direction, so all kymograph records should be read from left to right.

1.12 Smoked Paper Records

Moisten the gummed edge of the kymograph paper. Holding the gummed edge in the left hand wrap the paper, glossy side out, round the drum. When square and tight, press down the gummed edge. If the paper shows a tendency to slip on the drum, secure it with transparent plastic tape at the edge. The layer of carbon is applied to the paper on the drum in a fume cupboard. Coal gas is bubbled through a bottle containing benzene. In cold weather the benzene container should be put in a water bath at 25° C. This gives a very sooty flame when a fish tail burner is used. Rotate the drum rapidly and bring the flame slowly and steadily down the paper; only the yellow part of the flame should be in contact with the paper. The layer of soot should be dull black-brown, the paper underneath white and unscorched.

Time Marking

It is necessary to indicate the peripheral speed of the drum surface during a recording. The type of marker used depends on the drum speed; where time intervals of 1/100th second are necessary, an electromagnetic marker tuned to 100 Hz_1 is driven from the 50 Hz a.c. mains supply via a step-down transformer. The time-marker is mounted on a pivoted stand and its writing stylus is swung against the drum for a single revolution.

For slow drum speeds, time marks at intervals of 1, 5, 10 or 60 seconds are produced by an electromagnetic signal driven by timed pulses from a clock.

After the recording has been made on the smoked surface, cut the paper off the drum and lay it flat on the bench. Use a glass rod with a fine ball point at the end to write explanatory notes on the record. Varnish the paper by pulling the paper slowly once through a pool of varnish. Allow the excess varnish to drip back into the pool and then pin up the tracing on the drying rack. When dry, trim the records to the minimum size and stick them in your book close to the experimental instructions.

1.13 Conditions to Ensure the Survival of Isolated Tissues

Isolated surviving organs and tissues are used in many of the experiments in this book. They will remain alive for many hours if surrounded by the correct physical and chemical environment during experiments, i.e. if immersed in a solution containing the correct concentrations of salts, within the permitted range of temperature and pH and with an adequate oxygen supply. The simpler experiments use frog tissues; these work well at room temperature and enough oxygen will diffuse from the air into shallow solutions. Other experiments use tissues taken from the rat, rabbit and guinea-pig and a variety of solutions have been described to bathe these tissues. First some general comments will be made on the method of preparation of these solutions and then the composition of the solutions recommended for particular tissues will be given.

The water used to make up the solutions must be either distilled or deionized. The solutions can be kept in Pyrex glass containers and there is little point in making up small quantities. A 10 or 20 litre stoppered aspirator will store solutions satisfactorily for weeks provided that any glucose necessary is added just before use. The chemicals used should be of the purest quality obtainable, of the grade used for chemical analytical work; the one most likely to be troublesome is calcium chloride owing to doubts about the water content. It can be used by dissolving the contents of a fresh unopened bottle of the hydrated salt $CaCl_2.6H_2O$ in distilled water to form a standard solution, e.g. 1 g. $CaCl_2$ per 10 ml. of solution, checked if necessary by titration with standard $AgNO_3$ solution, which can then be dispensed volumetrically. An M/1 solution is available commercially[9]. It should be added to the mixture last, apart from the glucose, and into almost the final volume.

Frog Tissues

Ringer's Solution

NaCl 0.6 per cent., KCl 0.0075 per cent., $CaCl_2$ 0.01 per cent., $NaHCO_3$ 0.01 per cent.

Mammalian Tissues

HEART

Locke's Solution

NaCl 0.9 per cent., KCl 0.042 per cent., $CaCl_2$ 0.024 per cent., $NaHCO_3$ 0.015 per cent., glucose 0.1 per cent.

pH adjusted to 7.3 to 7.4

INTESTINE

Tyrode's Solution

NaCl 0.8 per cent., KCl 0.02 per cent., $CaCl_2$ 0.02 per cent., $NaHCO_3$ 0.1 per cent., $MgCl_2$ 0.01 per cent., NaH_2PO_4 0.005 per cent., glucose 0.1 per cent.

pH adjusted to 7.3 to 7.4

DIAPHRAGM

Krebs' Solution

NaCl 0.69 per cent., KCl 0.035 per cent., $CaCl_2$ 0.028 per cent., $NaHCO_3$ 0.21 per cent., KH_2PO_4 0.016 per cent., $MgSO_4.\ 7H_2O$ 0.029 per cent., glucose 0.2 per cent.

1.14 Drug Concentrations

Drugs are used in isolated organ experiments in very dilute solutions. It is usual to describe the concentration of the drug as the weight of the drug contained in 1 ml. of solution. Thus the concentration of adrenaline in a stock solution containing 1 mg. (.001g) per ml. is 10^{-3} or acetylcholine containing 2 μg (.000002 g) per ml. is 2×10^{-6}.

For accurate work there are advantages in expressing the concentration in molarity since this indicates the active mass. A drug is usually available in a number of forms, e.g. acetylcholine is supplied as the chloride, bromide and iodide; the chloride will appear more active than the iodide because of its lower molecular weight if the concentration is expressed on a weight rather than a molarity basis.

To convert weight concentration to molarity it is necessary to know the molecular weight of the compound and also if each molecule contains a single molecule of drug. Of the drugs used in this book only atropine sulphate and physostigmine sulphate have two molecules of drug per molecule of compound, the remainder have a single drug molecule. A list of molecular weights of the drugs used in organ bath experiments is given below.

Acetylcholine chloride	181·7
bromide	226·2
iodide	273·1
Adrenaline	183·2
hydrochloride	219·7
hydrogentartrate	333·3
Atropine Sulphate $\frac{694 \cdot 8}{2} =$	347·4
Neostigmine bromide	303·2
methane sulphonate	334·3
Oxytocin (1 mg. = 500 i.u.)	1006
Physostigmine sulphate $\frac{648 \cdot 8}{2} =$	324·4
salicylate	413·5
Procaine hydrochloride	272·8
(+)—Tubocurarine chloride $5H_2O$	785·8

A molar solution contains the molecular weight of the compound in grams dissolved in a litre of solution, e.g. if the molecular weight is 100 each litre of the molar solution contains 100 g. and each ml. contains 100 mg. which is a weight concentration of 10^{-1}.

To convert concentration expressed as g. per ml. to molarity multiply by the fraction $\frac{1000}{\text{mol. wt.}}$, to convert molarity to weight concentration multiply by the fraction $\frac{\text{mol. wt.}}{1000}$.

Example. Express a 10^{-6} solution of tubocurarine as a molar concentration.

$$10^{-6} \times \frac{1000}{785 \cdot 8} = 1 \cdot 25 \times 10^{-6}\text{M}$$

If the drug is dissolved in distilled water, it should be added to the organ bath in a relatively small volume (e.g. less than 5%) so as not to disturb the composition of the fluid bathing the tissue.

CHAPTER TWO

MUSCLE AND NERVE

2.0 Introduction

It is convenient to consider the physiology of muscle and nerve together. Nerve fibres control the activity of skeletal muscle fibres and the first experiments use isolated nerve-muscle preparations dissected from the frog. Frog tissue has the practical advantage that it will function at room temperature without a blood supply; its oxygen requirements are met by diffusion from the air into the solution bathing the preparation. The frog experiments have been arranged so that the properties of the nerve-muscle preparation are shown first, later experiments illustrate the links in the chain of events between the nerve impulse in the motor fibre and the development of tension in the muscle fibre.

There is a great range in the size and fibre arrangement of the muscles attached to the skeleton of the frog, traditionally the gastrocnemius muscle and sciatic nerve are used for the simpler experiments, supplemented by the sartorius and extensor digitorum longus IV. It should be remembered that the frog has two types of skeletal muscle, a twitch fibre similar to the mammalian twitch fibre and also a slow postural muscle which is not represented in the mammal with the possible exception of fibres in the extra-ocular muscles and intrafusal muscles.

The rectus abdominis muscle is used to demonstrate the activity of the slow fibres, all the other experiments are concerned with the fast twitch fibres.

Following the frog experiments, an isolated mammalian skeletal muscle preparation is described; the rat diaphragm, because of its large surface, can be supplied with oxygen by diffusion from the bath. Some experiments with human nerve and mammalian smooth muscle complete the chapter. Intestinal smooth muscle activity is described in Chapter 9.

2.1 The Frog Sciatic Nerve and Gastrocnemius Muscle Preparation

The frog is held by the legs, ventral surface uppermost and stunned by swinging its head sharply against the edge of the bench. Its head is cut off and the spinal cord destroyed by pushing a large pin or wire down the vertebral canal with rotary movements.

This procedure is called pithing. If frog heart preparations are in demand at the same time it is economical at this stage to divide the frog into an upper and lower part. Hold the pithed frog ventral surface uppermost, divide across the abdomen so as to leave small piece of vetebral column in the lower half. The upper part is available for heart experiments, two gastrocnemius preparations can be made from the lower part.

During dissection do not allow the tissues to dry; exposed nerve and muscle should be flooded with Ringer's solution. Try not to touch nerve tissue at all, pick it up by holding adjacent connective tissue, do not stretch it and, of course, never compress it with forceps. A glass rod with a small bead at the end can be used to lift a nerve from beneath.

Remove the skin from the lower part of the frog, either with scissors or pulling it off from the legs in one piece.

Lay the frog on a cork board dorsal surface uppermost. Compare with Figure 2.1 and identify the muscles. Cut out the urostyle (the prolongation of the vertebral column) with scissors by freeing its distal end and cutting the muscles on its lateral margins, detach from the spinal cord (section 1 in Fig. 2.1). You will see the sciatic nerves running into the vertebral column. Use scissors to divide the vertebrae in the midline (2) cut the connection to the ileum (3). Separate the large muscles of the thigh by cutting the connective tissue (4), the ileofibularis muscle will be seen between them, and beneath lies the sciatic nerve trunk. Lift up the fragment of vertebrae with sciatic nerve attached and dissect the nerve free as far as the knee joint cutting the small nerve branches which leave the nerve at hip joint and those that supply thigh muscles.

Cut through the plantar insertion of the gastrocnemius distal to the sesamoid bone (5). Separate the gastrocnemius muscle from the tibio-fibula, cutting the latter together with the anterior tibial muscles, near the knee joint.

Cut through the femur and thigh muscles close to the knee (6) being careful to avoid the sciatic nerve. Put the preparation in a dish containing Ringer's solution.

In the following experiments the sciatic nerve is often exposed to air and may begin to dry. Spontaneous muscle movements due to the initiation of nerve impulses in the drying nerve are a warning sign to immediately replace the nerve in Ringer's solution.

2.2 Frog Muscle Contraction

Connect the apparatus as shown in Figure 2.2. The circuit is arranged so that the stimulator is triggered to emit one stimulus each time the contact-maker is struck by the rotating arms on the kymograph spindle. Fill the muscle bath with Ringer's solution. Tie a thread round the tendon of the gastrocnemius muscle and secure the other end, at a distance of about 3 cm to the short arm of the recording lever. This limb dips into the fluid in the bath. Push a pin through the knee joint or cavity of the femur. Insert the protruding tip of the pin into the socket on the floor of the frog bath. Adjust the position of the lever on the side of the bath so that the long arm is horizontal when its weight is borne by the muscle, i.e. with the afterloading screw not making contact with the lever. The muscle is now said to be free-loaded or free-weighted. Screw up the afterloading

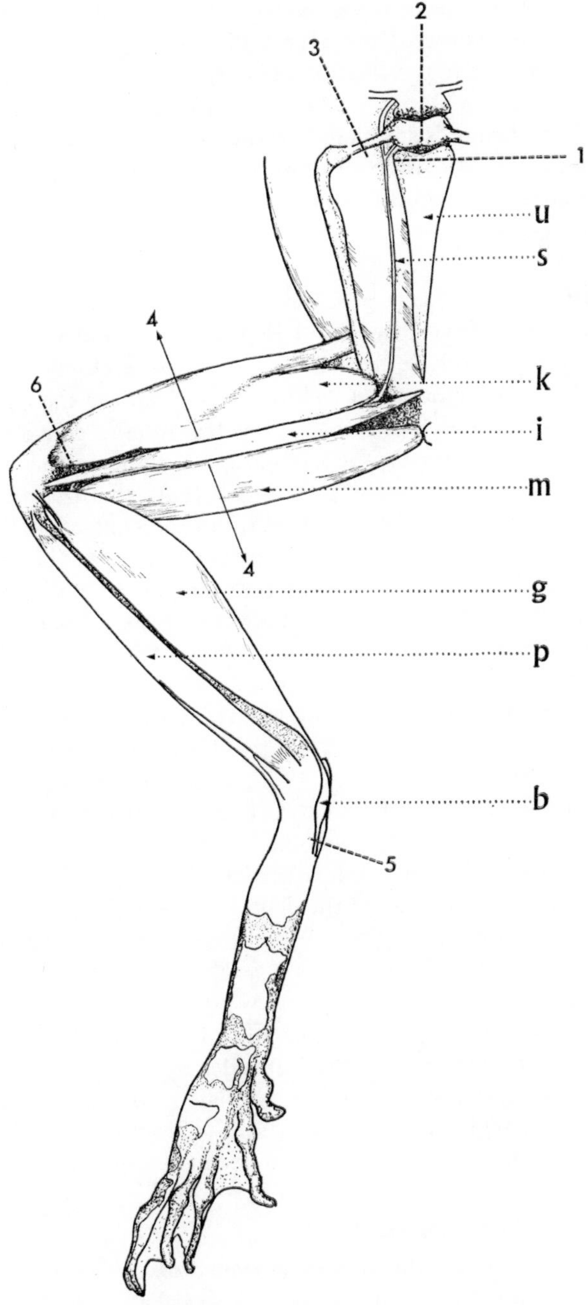

Fig. 2.1

Dorsal view, the numbers refer to instructions in para. 2.1. Letters are as follows, u, urostyle: s, sciatic nerve: k, gluteus muscle: i, ileofibularis muscle: m, semi-membranosus muscle: g, gastrocnemius muscle: p, peroneal muscle: b, sesamoid bone.

screw until the weight of the lever is just taken off the muscle. Adjust the electrodes so that they are a few millimetres apart and a few millimetres above the surface of the liquid. Lay the nerve across them, and put the fragment of spinal cord at the end of the nerve on the edge of the bath so that the nerve is in air. Make sure that the stimulus will not be short-circuited through Ringer's solution. The nerve must not be allowed to dry, so only leave it on the electrodes for the time necessary to carry out the experiment. If it begins to dry it may generate nerve impulses spontaneously, and the muscle will twitch irregularly and spoil the recording.

Stimulator Instructions. Turn Selector switch to PULSE OFF, Strength to 0, and Pulse Rate to 5 per sec. Move Selector to REPEAT PULSE and advance Strength until maximal contractions of the muscle occur at the rate indicated by the neon indicator lamp on the stimulator. Move Selector switch to PULSE OFF. Connect the EXT. TRIG. terminals to the kymograph contact-maker terminals. Set the spindle contact arms 180° apart. Adjust the position of the drum on the kymograph spindle so that the records of the twitches will not occur across the join of the paper. Remember that the beginning of the contraction occurs soon after the contact-arm strikes the contact-maker. Set the drum rotating at about 25 cm per second. Move Selector to EXT. TRIG. for a few drum rotations and ensure that the muscle contracts twice per revolution. Return Selector switch to PULSE OFF. Now adjust the position of the recording lever on the drum surface so that the pointer is tangential to the surface, horizontal, and pressing lightly on the smoked surface. Start the drum rotating, wait two seconds, and switch Selector to EXT. TRIG. for a single rotation, i.e. to record two twitches. Inspect the traces; if the top of the record is missing, firmer pressure of the pointer on the drum or adjustment of the levelling screws is required. Now mark the point on the record at which the stimulator was triggered. To do this turn Selector to EXT. TRIG., rotate the drum slowly by hand in the same direction as before and slowly operate the contact-maker with the spindle arm. The muscle will contract and draw a single vertical line on the drum. Repeat the procedure for the trace on the other side of the drum. Move the lever away from the drum surface and replace the nerve in the Ringer's solution. Set the drum rotating and bring the 100 Hz time marker against the drum for a single rotation, at a level so that it writes just below the baseline of the twitch record. Measure the latent period, i.e. the time interval between the moment when the stimulus was applied to the nerve and the beginning of the contraction of the muscle. Consider the events which have occurred in this interval. Examine the lever system and calculate (a) the shortening of the muscle which produced the recorded twitch (b) the mass lifted by the muscle (c) the work done by the muscle. 'Work done' equals force times distance. See Figure 2.3.

2.3 Isometric Muscle Contraction

In the preceding experiment the muscle shortened against a force which did not vary very much during the course of contraction—this is called isotonic contraction. The record displays change of muscle length with respect to time. If the muscle is secured to

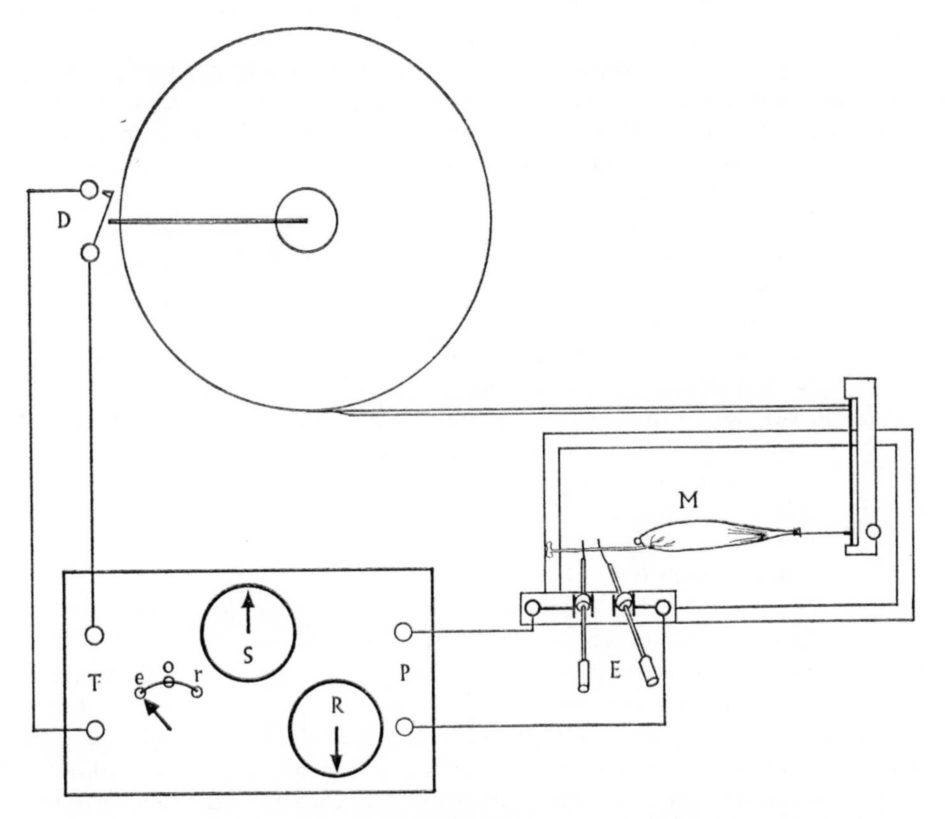

Fig. 2.2

A sciatic nerve and gastrocnemius muscle preparation (M) is arranged in the organ bath with electrodes (E) supporting the nerve. The electrodes are connected to the output terminals (P) of the stimulator. The positions of the selector switch are shown, namely External Trigger (e), Pulse Off (o) and Repeat Pulse (r). Strength control is (S) and repetition rate is (R). The external trigger terminals (T) are connected to the contact-maker (D) on the kymograph. The SRI_{33} assembly incorporates the stimulator in the base of the kymograph, so connections to the contact breaker are established by switches rather than external wiring. An inkwriting stylus is used and the recommended drum paper carries pre-printed time calibration lines. Generally speaking the type of glazed paper prepared for smoked drums is not suitable for inkwriting styluses, and vice versa.

rigid supports at either end so that it cannot shorten when active, then it generates tension, and this is called isometric contraction.

The simple method of directly recording isometric contraction employs a lever system with a stiff metal spring; the general arrangement of the apparatus is as in paragraph 2.2, however note that some shortening must occur in order to move the lever, but because the lever movement is proportional to tension applied by the tendon, the record is the change of tension with respect to time.

A better method uses a force transducer and displays the time tension graph on the face of an oscilloscope. Use a double pulse stimulator to trigger the oscilloscope and set the controls as below.

Oscilloscope Trigger—external
Time-base velocity—20 msec/cm
Y amp. sensitivity—20 mV/cm

Stimulator Insert jack with push button into single-shot jack socket so that you can initiate stimulations manually.
Delay C1 > 1 msec
Width C1 midway
Strength C1 range 0–18v dial 6
Output switch C1
Output jack in C1—jack connected to stimulating electrodes
External trig. sockets to EXT. TRIG. of oscilloscope

Check that pressing the push button initiates a sweep on the time-base and a twitch in the muscle. The tension transducer (see p. 6) is mounted on a rackwork so that it can vary the resting length of the muscle. Adjust this until it is moderately extended, at approximately its natural length in the frog. Figure 2.4 shows a tension recording under isometric conditions. Compare time relations of this with a contraction record under isotonic conditions. Consider the situations in which you use your muscles isometrically and isotonically.

2.4 Effect of Temperature on the Course of Muscular Contraction

Connect EXT. TRIG. to drum contacts. Speed, about 25 cm per sec. Take two twitches (180° apart) as described before (2.2). Swing the writing point off the drum, pour Ringer's solution at 0° C into the muscle bath, leave for 2 minutes and refill with more solution at 0° C. Replace the lever on the drum and take another trace (i.e. two twitches) on top of the previous trace as soon as possible. Swing the lever off and fill the bath with Ringer's solution at 27° C; record two further twitches on top of the previous ones. Compare the forms of the three twitches which have, of course, the same moment of stimulation. Measure the latent periods.

2.5 Effect of Prolonged Exercise (Fatigue)

Since the pressure of the writing point on the drum surface affects the size of the twitch, it is necessary in this experiment to apply the writing point to the drum with the same pressure throughout the experiment. To do this slacken off the clamping screw and

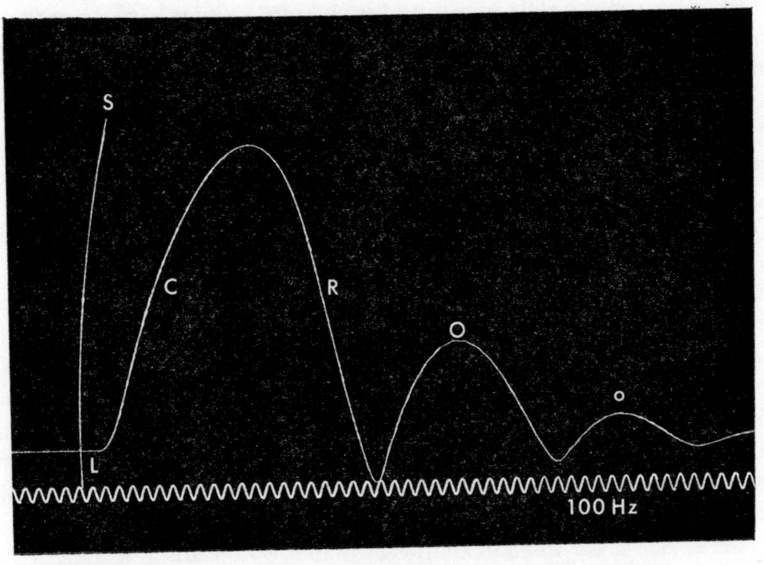

FIG. 2.3

Smoked drum recording of the isotonic contraction of a frog gastrocnemius muscle. (S) is the moment of stimulation, (L) is the latent period, (C) the contraction period, (R) the relaxation period, (O) and (o) are oscillations of the lever system. Compare with the isometric record 2.4.

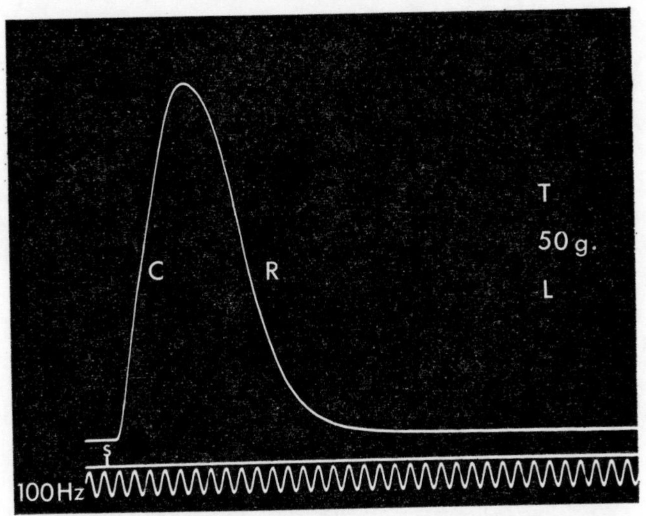

FIG. 2.4

Tension developed during isometric contraction of a frog gastrocnemius muscle. The tension was detected by a strain gauge transducer and displayed on a cathode-ray oscillograph. Compare the time course of the development of tension with the shortening process shown in the isotonic record in Fig. 2.3. Stimulus marked by (s).

rotate the slotted base-piece clockwise until the end of the slot is hard against the clamping screw; move the main base-piece into position so that the lever writes properly on the drum. The writing point can now be swung off the drum by rotating the slotted base-piece anticlockwise without moving the main-base piece and reapplied by reversing the rotation until the end of the slot touches the clamping screw. Connect the drum contact-maker contacts to the Ext. Trig. terminals. Set Strength control to give maximal contractions. Record two twitches (180° apart) at drum speed of about 25 cm per second. Mark the moments of stimulation. Swing the lever off the drum, start the drum once more. With each revolution of the drum the preparation will be stimulated twice but without making a record. Count the twitches '3, 4, . . . 19, 20'. Swing the lever smartly into position as described above and record 21, 22. Swing the lever off again and continue counting '23, . . . 40'. Record 41, 42, and so on until fatigue is quite marked. With a good preparation it may be better to record 1, 2; 51, 52; 101, 102, etc. When the muscle-nerve preparation is completely fatigued place the electrodes on the muscle and stimulate again. Allow 15 minutes for recovery and stimulate the nerve again. Record a 100 Hz time trace below the trace. Measure changes in the latent period, contraction amplitude and contraction rate that occur during fatigue.

This experiment can be done in another way to give a continuous fatigue curve. Use a peripheral drum speed of 1 to 1·5 mm per sec. Set Rate to 1 per second, Selector to Repeat Pulse and Strength to a level to give maximum contractions. Record all the contractions against the slowly rotating drum and continue until the muscle is completely fatigued. The effect of rest pauses can be shown by this method.

2.6 Effect of Load and Length on Muscle

The apparatus required is the same as for paragraph 2.2. Connect the output of the stimulator to the electrodes but do not connect Ext. Trig. to the kymograph contact-maker. Put Selector switch at Pulse Off position and Rate at 1 per second. Arrange the muscle so that the recording lever is horizontal and just supported by the afterloading screw with no weight on the lever. The muscle will thus not have to work against the weight until it begins to shorten—i.e. it is afterloaded. Another way of describing this is to say that the series elastic element in the muscle is not, at rest, extended by the load. Put the Selector switch to Repeat Pulse and advance Strength until a maximal twitch is obtained. Return Selector switch to Pulse Off. Bring the writing point lightly against the drum about half way down it. Take a record—it will be a single line—of the contraction of the unloaded muscle on the stationary drum by moving the Selector switch to Repeat Pulse for a single pulse. Move the drum round by about 0·5 cm by hand. Place a 5 g. weight on the serrated edge of the muscle lever. Put a pencil mark on the lever at this position. Record a contraction, again move round the drum by about 0·5 cm. Continue to add weights, advancing by 5 g. on each occasion and always placing the weight on the serrated edge of the lever at the same place. Take records until a weight is reached which the muscle is just unable to lift. Pay attention to the speed with which the lever rises during the twitch. It will move more slowly as the load is increased. This is called the Force-Velocity relationship. Move the drum on a further 0·5 cm, lower the afterloading

screw and allow the muscle to be stretched by this heavy weight without support from the screw. The muscle is thus loaded at rest, the series elastic element and the contractile element in the muscle will be extended. Move the drum along another 0·5 cm. Stimulate the nerve again; the muscle will contract and raise the weight. Remove the weights. Label the tracing with figures indicating the weights. See Figure 2.5 for an example.

Calculate the work done at each twitch and draw a graph relating work (ordinate) and load. Write a note, from your own experiences, of the relationship between rate of muscular contraction and load lifted.

2.7 The Length-Tension Relationship

(*a*) *Direct Recording.* A vertical gastrocnemius muscle isometric assembly (Palmer type C205) is used. The adjustable stand enables the resting length of the muscle to be varied, and the isometric tension developed is recorded on a smoked drum. Electrodes allow the stimulus to be applied either via the sciatic nerve or directly to the muscle. Fit a manual switch to the stimulator and use maximal stimulus strength. The drum is stationary during the contraction, so that a short vertical line is drawn, and then moved on by two or three mm for the next contraction. Extend the muscle by increments of 0·5 mm, record the resting and active tension at each length. A calibration weight is provided to record the sensitivity of the system; hang this on the lever at the same point that the muscle tendon is secured and measure the deflexion. Produce a graph relating muscle length to active and resting tension.

(*b*) *Electrical Recording.* A force transducer$_{11}$ is used to provide a signal to the oscilloscope. The general arrangement is as described in section 2.3. An isometric tension curve for the muscle twitch is recorded for each 1 mm increment of muscle length. A storage oscilloscope$_{4,\ 5}$ is ideal to display the family of curves produced, otherwise the traces may be recorded photographically$_{30}$. An alternative arrangement is to switch off the oscilloscope time-base and apply to the X plate amplifier the output from a position transducer (such as a multiturn potentiometer or linear potentiometer) attached to the rackwork mechanism which moves the force transducer. Thus the X axis now indicates muscle length and the Y axis muscle tension. Moving the length rackwork will trace out an XY curve of the length-tension relationship both for active and resting tension. Figure 2.6 shows an example of this type of display. Produce a graph relating muscle length to active and resting tension.

2.8 The Summation of Two Contractions

Set up a muscle bath and stimulator as in paragraph 2.2. Separate the two contact-maker arms on the spindle of the kymograph by 45°. Set drum speed at 25 cm per second. Since this corresponds to about 2 seconds per revolution, the lagging arm will strike the contact-maker about 0·25 sec after the leading arm. Operate stimulator in EXT. TRIG. mode and record two muscle twitches, together with moments of stimulation. Lower the drum, and move the lagging arm nearer the leading arm so that the second stimulation occurs during the relaxation phase of the leading twitch. Record the summed

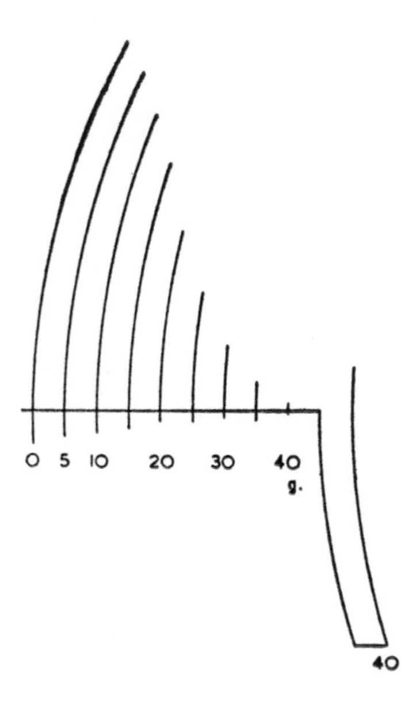

FIG. 2.5
The effect of load and length.

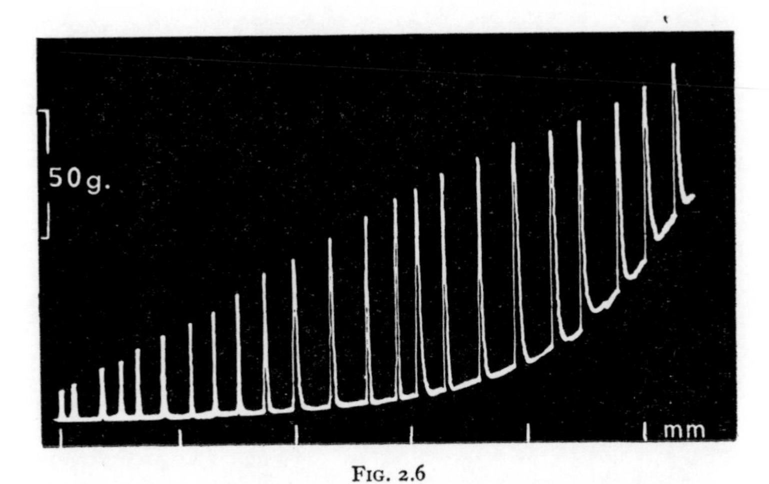

FIG. 2.6
The length-tension relationship of a frog gastrocnemius muscle. The muscle length is coupled to the X axis of the cathode-ray oscillograph. tension developed at rest and by the twitch is shown on the Y axis.

contraction. Move the arms closer so that the second stimulation occurs during the contraction phase of the first twitch. Record a 100 Hz time trace and mark all moments of stimulation.

2.9 The Production of a Sustained Contraction

Twitch contractions rarely occur under natural conditions; the usual method is to produce shortening or tension in the muscle by the discharge of a series of nerve impulses. To show the effect of the frequency of discharge on the contraction, set the Selector switch to PULSE OFF, the Strength to a level that will give maximal contraction and the Rate to 1 per sec. Use a slowly moving drum (4 mm per sec). Apply the lever to the drum, move the Selector switch to REPEAT PULSE and record for five seconds. Repeat with the Pulse Rate set at 5, 10, 15, 25 and 50. Write the stimulation rate under the appropriate section of the record. When your partner repeats this experiment, study the behaviour of the muscle at the different rates of stimulation.

2.10 The Frog Sartorius Muscle

Two sartorius muscle preparations are used; one is the muscle alone, this is easily dissected in a few minutes, the other has a length of nerve attached, this dissection may take an hour and requires skill and familiarity with the arrangement of the thigh muscles. Good lighting and a binocular magnifier are necessary; for the initial parts of the dissection a magnification of $\times$ 4 is convenient, for the nerve dissection a magnification of $\times$ 10 or $\times$ 15.

Dissection of the muscle. Stun, decerebrate and pith a frog. Lay it ventral surface uppermost on a cork board. Remove the skin from the lower abdomen and thigh. The position of the sartorius muscle is shown in Figure 2.7, it emerges from beneath the rectus abdominis muscle and lies superficially on the medial surface of the thigh, terminating distally with a short slender tendon attached to the knee.

Use a mounted needle to slit the connective tissue at the margins of the muscle near the knee. Extend this dissection distally to clear the tendon, pass a thread underneath, and tie tightly, taking care not to crush any muscle fibres. Detach the tendon, lift up the end of the muscle and dissect it free towards the pelvis. In mid-thigh you will see a blood vessel and nerve trunk entering the inner surface of the muscle; cut these. Detach the rectus abdominis muscle to reveal the attachment of the sartorius to the pubis. Press down with the blade of a scalpel to divide the pubis symphysis slightly off the midline away from the side you have dissected the sartorius. Undercut the bone beneath the muscle attachment. You now have a thin, slightly triangular muscle with a fragment of bone at the central end and a thread attached to the tendon at the distal end. Immerse the muscle in a dish of Ringer's solution.

The Sartorius Muscle-Sciatic Nerve Preparation. Follow the preceding instructions but with the following modifications. Expose the muscle and tie a thread on the tendon. Identify the position of entry of the nerve into the muscle. The nerve passes inwards and dorsally between the margins of the gracilis and the adductor magnus muscles. Detach the gracilis from the knee, separate it from the sartorius muscle, and remove entirely. Remove the semitendinosus tendon from the knee and peel back the muscle. Turn the frog over and detach the gluteus and ileofibularis muscles from their pelvic attachments. The main trunk of the sciatic nerve will now be visible. Detach and remove the semimembranosus muscle. You will now see a slender branch of the sciatic nerve running across the deep surface of the adductor muscle to the sartorius muscle. Section the branch where it joins the sciatic and dissect it back to the sartorius. Turn the frog over and remove the sartorius complete with nerve.

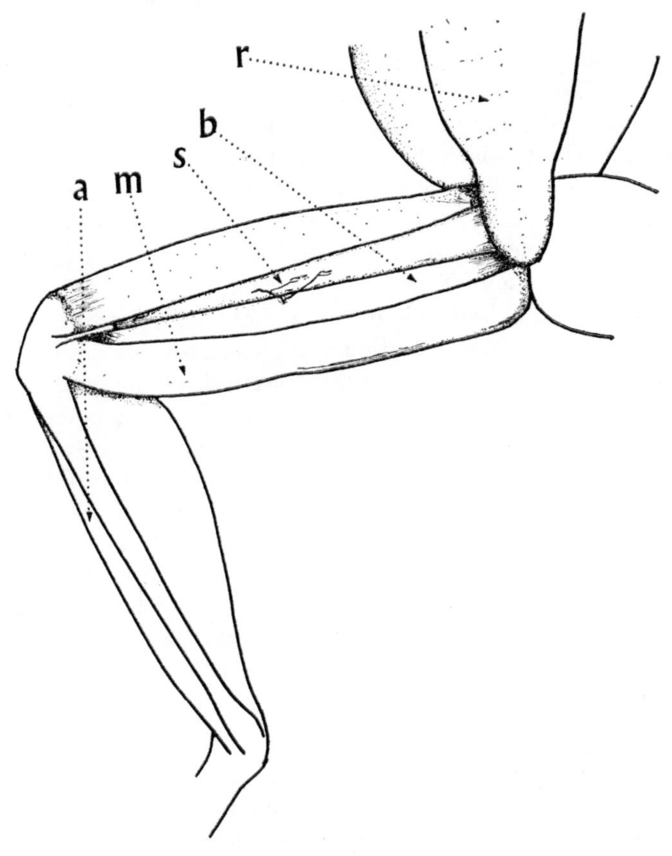

FIG. 2.7

Ventral view. The dissection is described in para. 2.10. Muscles indicated are as follows, a, anterior tibial muscle: m, gracilis muscle: s, sartorius muscle: b, adductor magnus muscle: r, rectus abdominis muscle. See also Fig. 2.1.

2.11 Conduction Velocity of the Muscle Action Potential

Dissect out a frog sartorius muscle as described in section 2.10. The arrangement of the muscle bath is shown in Figure 2.8. The muscle is extended to about its resting length and held by plastic plates or mounted forceps. The inner surface of the muscle, which is relatively free of connective tissue, should face uppermost in the bath. Adjust the stimulating electrodes to touch the surface of the pelvic end of the muscle. Suck out the Ringer's solution and replace with paraffin oil. Set the stimulator (2CROS) controls at, Frequency, 2 per sec; Delay, < 1 msec; Width, intermediate; Strength, range 0 to 18v; Position, zero. Set the oscilloscope controls at, trigger, external from stimulator; sweep speed, 2msec/cm; overall Y sensitivity, 1mv/cm.

Watch the muscle through the microscope and gradually increase the strength of the stimulus until the fibres nearest the stimulating electrode are seen to twitch twice a second. Now adjust the position of the pickup electrodes to lie over the active fibres at the far end (knee joint tendon) of the muscle. An irregular action potential should appear on the oscilloscope. Observe how the shape is affected by the strength of stimulus. Measure the conduction time of the fastest fibres, and move the pickup electrodes 10, and 20 mm towards the stimulating electrodes, again measuring the conduction time. Calculate the velocity of the action potential in metres per second.

The shape of the recorded action potential is influenced by the separation of the pick-up electrodes. Set the stimulus strength at a value at which only one or two muscle fibres are excited and compare the action potential when the inter-electrode separation is 15 mm, 10 mm and 1 mm. Note that a biphasic action potential with isopotential interval is produced with large electrode separations.

2.12 The End-Plate Potential

Dissect out a frog sartorius muscle with about 1 cm of nerve attached and set up in Ringer's solution in a small horizontal Perspex bath as shown in Figure 2.8. Use bipolar electrodes within the paraffin layer to stimulate the nerve. The stimulator triggers the time-base of the oscilloscope with a sweep speed of 1 msec/cm. Bipolar pick-up electrodes are mounted on a manipulator so they can traverse the length of the muscle; the traverse mechanism is graduated in mm. Focus the binocular microscope (× 10 or × 20) on the surface of the muscle and make a drawing of the course of the larger nerve trunks.

Adjust the level of the Ringer's in the bath so that the upper surface of the muscle is in contact with the paraffin oil. Stimulate the nerve at a frequency of 1 shock per second, note the contractions visually and record muscle action potentials from the surface of the muscle. Measure on the oscilloscope the time delays between stimulus and the earliest muscle action potential at different positions along the length of the muscle. The type of record obtained is quite different from that obtained in experiment 2.11 because the stimulus is being transferred rapidly by nerve fibres to several end-plate regions, thus the minimum latency should be obtained when the electrodes are over an end-plate zone; conduction by muscle fibre, which is relatively slow (2 or 3 mm per

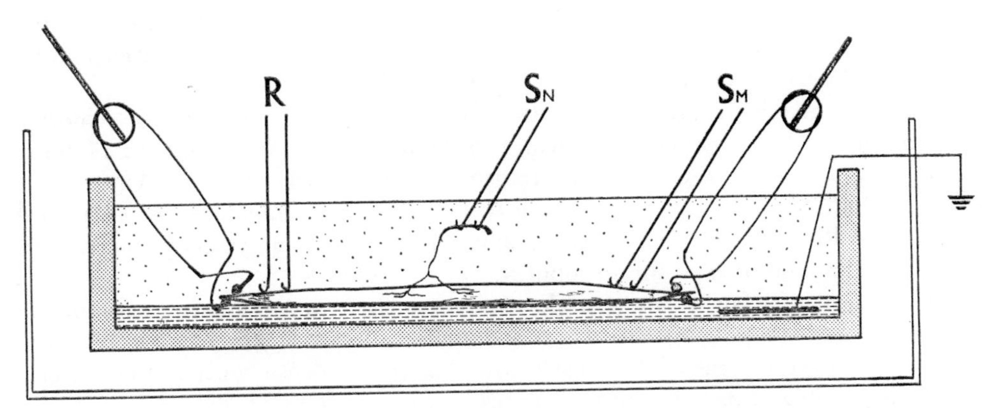

FIG. 2.8

The frog sartorius muscle is shown extended between spring clips in a bath. The muscle lies at the interface between a shallow layer of Ringer's solution below and oil above. Recording electrodes, R, dip into the oil layer and touch the surface of the muscle. The nerve can be stimulated by electrodes SN in the oil layer. The muscle can be stimulated by electrodes SM in the oil layer. An earthing plate is immersed in the saline layer and the bath is surrounded by a metal screen. In experiment 2.11 electrodes SM and R are required, in experiment 2.12 electrodes SN and R are used.

msec), to the pickup electrodes will increase the latency. Draw a graph relating position of the pickup electrodes i.e. mm from tendon to latency. From the graduations on the manipulator locate the positions of minimum latency on the muscle length.

Suck out the bath and refill with Ringer containing tubocurarine chloride, concentration 10^{-6}M (1 μg per ml.). Change the bath contents with fresh curare solution once or twice. Test the neuromuscular transmission every few minutes. When neuromuscular block, as evidenced by failure to twitch on nerve stimulation, has developed, suck out the Ringer and replace with paraffin oil. It may be necessary to increase the strength or duration of exposure to tubocurarine chloride.

Traverse the surface of the muscle with the pickup electrodes while stimulating at 10 shocks per second. Muscle action potentials will now be absent or infrequently elicited, but in the regions of the muscle where there is a dense concentration of end-plates it is possible to pick up the non-propagated compound end-plate potential. See Figure 2.9, note the increased Y plate sensitivity required to detect the end-plate potential. As the bipolar pick-electrodes are moved towards an end-plate region the deflexion should increase in magnitude, then on passing away on the other side the potential reverses polarity and declines. Measure the time-course of the EPP and compare with a muscle action potential.

Do the end-plate regions correspond to the regions of minimum latency located in the first part of the experiment?

2.13 The Muscle Action Potential and the Development of Tension

Use either a frog sciatic nerve-gastrocnemius muscle preparation or innervated sartorius muscle. Arrange stimulator, tension transducer and oscilloscope as described in paragraph 2.3 but in addition apply silver wire or saline wick electrodes to the surface of the muscle. The muscle should be out of the Ringer's bath during the recording; if it is immersed the electrical size of the muscle action potential will be very much reduced by the short-circuiting effect of the Ringer's solution.

The output of the tension transducer should be applied to the Y1 beam and the output of the muscle action potential amplifier applied to the Y2 beam. Operate the stimulator with a push button in the single-shock mode. The shock should be maximal and the time-base velocity 5 msec/cm.

Apply single shocks and observe the relationship between the muscle action potential and the development of tension. To retain the image long enough to make measurements use either a long persistence oscilloscope, a storage oscilloscope or photography. Measure the time delay between the beginning of the muscle action potential and the beginning of the increase in tension. Figure 2.10 shows a record from a frog gastrocnemius muscle.

2.14 The Twitch-Tetanus Tension Ratio

The apparatus is arranged as in paragraph 2.13, time-base velocity 40 msec/cm. Operate the stimulator in the single stimulus mode by inserting a jack with push button attached into the 'single' external trigger socket.

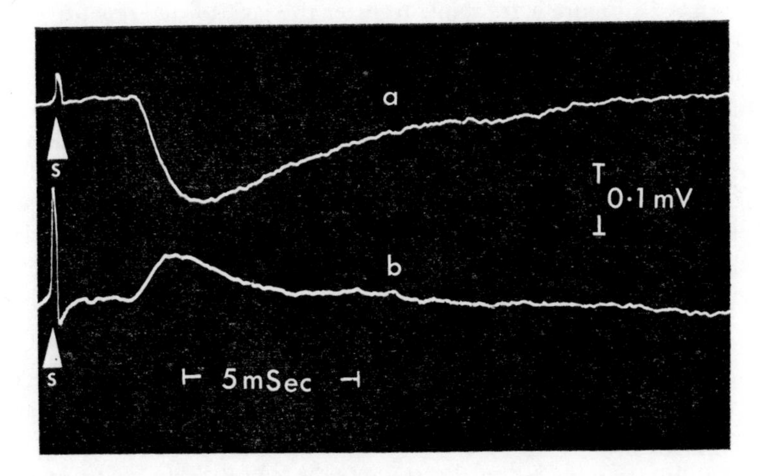

FIG. 2.9

The end-plate potential of a curarized frog muscle. The nerve was stimulated at point S, the electrodes were moved between traces (a) and (b) to show reversal of polarity.

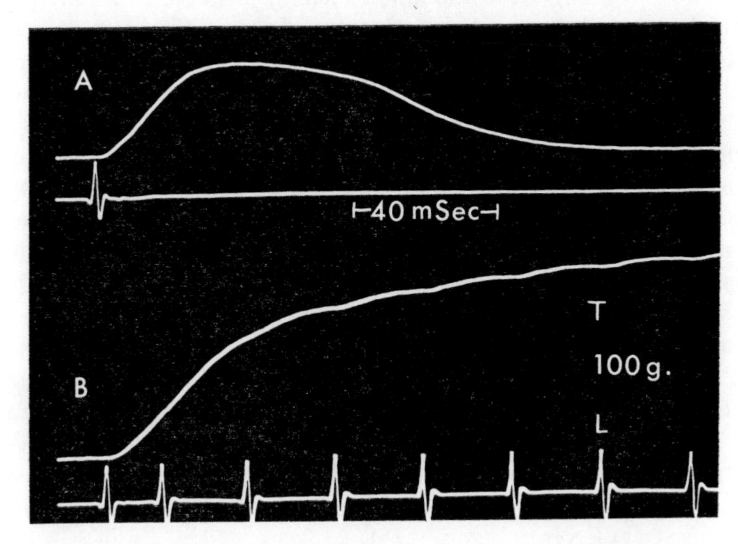

FIG. 2.10

Photographs from a double-beam cathode-ray oscillograph. Upper pair A show the response in a frog gastrocnemius muscle to a single shock, the action potential is shown on the lower trace, the tension development on the upper. Lower pair B, show response during a tetanus at 50 shocks per second.

Record an isometric tension response to a single maximal shock to the nerve, as shown in record A in Figure 2.10. Now transfer the jack to the 'repetitive trigger' jack socket and set the frequency at 50. Arranged in this way the stimulator will emit shocks at a repetition rate of 50 per second so long as the push button is depressed. Depress the push button, this will start the time-base, and hold it down for the duration of the time-base sweep. See record B in Figure 2.10. Notice that a larger tension is developed when the muscle is stimulated by a rapid series of shocks than when excited by a single shock. The ratio of the two tensions is called the Twitch-Tetanus ratio and it depends on the amount of elastic material between the contractile mechanism in the muscle fibres and the recording device. Read up about the Series and Parallel elastic elements of muscle in your text book.

2.15 Conduction Velocity of the Nerve Impulse

The apparatus is arranged so that the time interval of a few milliseconds which elapses between the stimulation of one end of a frog nerve trunk and the arrival of the compound action potential at the other end can be displayed on an oscilloscope. The synchronizing pulse from the stimulator triggers the time-base (sweep speed, 1 msec/cm) and 1 msec later the stimulator emits a square wave stimulus to the nerve. Bipolar electrodes pick up the compound action potential at the other end of the nerve, and after amplification it is displayed on the Y axis of the oscilloscope (overall Y sensitivity 1 mV/cm). The general arrangement of the apparatus is shown in Figure 2.11.

Dissect out from a large frog the longest length of nerve trunk available, i.e. start from the spinal cord and dissect the sciatic nerve down to the gastrocnemius muscle as given in instructions in paragraph 2.1, but continue the dissection of the peroneal branch to the ankle. Lay out the nerve on the millimeter scale of a plastic ruler and measure its length. Transfer the nerve to the bath, which consists of a plastic hemicylinder within a metal screening box. At each end are bipolar silver wire electrodes. An earth electrode consisting of a piece of silver foil lies in the bottom of the bath. Place the nerve in a shallow layer of Ringer and cover this with a layer of paraffin oil. Lift up each end of the nerve on to the electrodes, which are in the oil layer. The earth electrode and nerve trunk are in Ringer. The end of the peroneal branch should be crushed with forceps and applied to the distal pick-up electrode so as to give a monophasic action potential. Set the stimulator repetition rate at 10 per sec, delay at 1 msec, pulse width $\frac{1}{2}$ msec and increase the pulse height gradually from zero to 18 volts. A compound action potential will appear on the time-base and gradually change shape as the strength of the stimulus is increased. This is because the fibres in the nerve trunk are of different sizes, thresholds and conduction velocities, usually the largest, fastest fibres are stimulated first and the shape of the compound action potential changes as the slower fibres are brought in at greater stimulus strengths (see Figure 2.12).

The total length of nerve between the stimulating and recording electrodes has an effect on the shape of the compound action potential, since the greater the length, the greater will be the separation of the action potentials due to fibre groups of different conduction velocity.

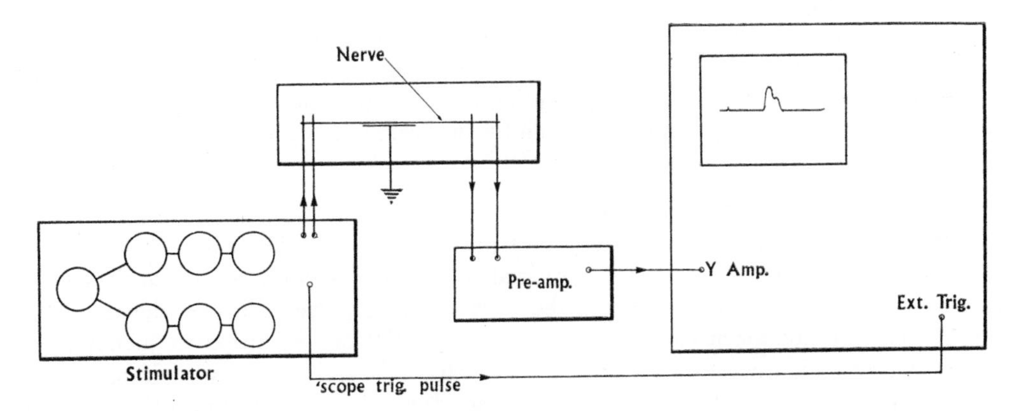

FIG. 2.11

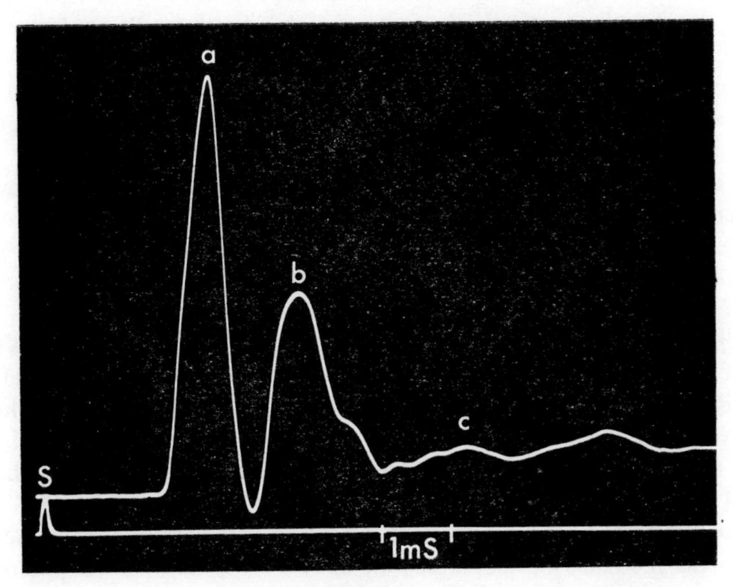

FIG. 2.12

A photograph from a cathode-ray tube of the compound action potential of the frog sciatic and peroneal nerve, total length 78 mm. The lower trace shows the stimulus (s) and the 1 msec time calibration. The first deflection (a) corresponds to a conduction velocity of 43m/sec, the second (b) to 25m/sec and (c) to 13m/sec.

Make some sketches, or take photographs of the compound action potential at different stimulus strengths.

Measure the time interval between the stimulus, as indicated by the shock artifact or the signal on the second beam of the oscilloscope, and the beginning of the major deflexions of the compound action potential, calculate the conduction velocity of each group by dividing the nerve length in mm by the conduction time in msec.

2.16 Refractory Period of Nerve

Set up the sciatic nerve and peroneal branch in the nerve bath as described in paragraph 2.15. The stimulator is arranged to give two stimulating square waves of independent size and delay following the synchronizing pulse. Further details of the double pulse stimulator are given in paragraph 1.3. Set the repetition frequency to 10 cycles per second, put pulse output on both channels at zero, pulse delays, 1 msec channel 1, 6 msec channel 2, pulse width $\frac{1}{2}$ msec. Increase the strength (pulse height) of channel 1 until a compound action potential appears on the time-base. Now increase the strength of channel 2, a second compound action potential will appear further down the time-base.

The stimulation times can be shown on the second beam of the oscilloscope directly below the action potentials, or may appear as shock artifacts. Vary the position of the second compound action potential by reducing the delay on channel 2 and note the interaction.

The nerve fibres are inexcitable, i.e. refractory, for a short period of time following a stimulus: can you tell from the display how long this refractory phase is? A sequence of records is shown in Figure 2.13.

Adjust channel 1 so that only the low threshold fastest fibres are stimulated. Set channel 2 strength at maximum. Since the slower, high threshold fibres are not stimulated by the first shock, the effect on the shape of the compound action potential of channel 2 as it is brought closer to channel 1 action potential should be limited to reduction of amplitude of the fast fibre component.

Immediately a nerve fibre has been stimulated to produce a nerve impulse it becomes refractory or inexcitable to further stimuli however large. This is the absolutely refractory period. This period is followed by a phase of reduced excitability during which the size of stimulus required to start a nerve impulse is abnormally high and the action potential produced is smaller than normal: this is the relatively refractory period. A phase of lower than normal threshold follows, this is called the supernormal period.

Set the pulse height controls of channel 1 and 2 at the threshold for the fast lowest threshold fibres. You should have a display of two compound action potentials on the time-base separated by an interval of 5 msec.

Reduce the delay on channel 2. When the compound action potential of channel 2 enters the relative refractory period of the preceding stimulus the size of the action potential will decline and then vanish. It can be restored by increasing the size of channel 2 stimulus (see Figure 2.14).

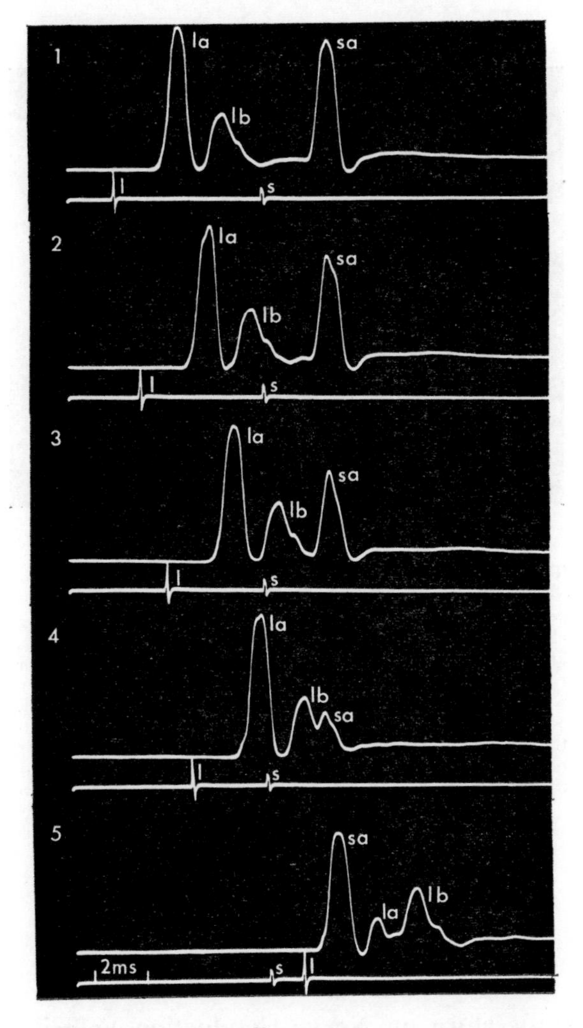

FIG. 2.13

Photographs of a double-beam cathode-ray tube showing five sweeps, numbered 1 to 5. A frog sciatic nerve was stimulated by a large shock (l) and a small shock (s) whose relative time positions were recorded on the lower of each pair of traces. The large shock excited both fast (la) and slower fibres (lb) of the compound action potential, the smaller shock excited only the fast (sa) deflection.

The time interval between l and s was decreased in traces 1 to 4 with l leading s, and in trace 5, s leads l. Note how the sa deflection was reduced as the relative refractory period of la invaded it in traces 1 to 4. In trace 5, sa nearly obliterated la. The lb deflection was unaltered throughout since these fibres were not stimulated by s.

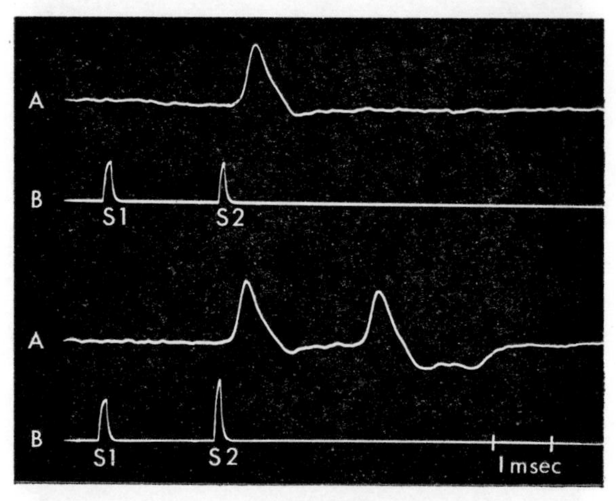

FIG. 2.14

Traces A and B were photographed from the face of a double-beam cathode-ray tube. Line A was the compound action potential picked up from the sciatic nerve following electrical stimulation. The points of stimulation, s1 and s2, are shown by the deflections on line B. In the upper pair of records, two shocks of equal strength separated by 2 msec were delivered to the nerve, but only a single compound action potential appeared because the second shock fell in the relatively refractory period of the nerve. In the lower pair of records the timing of the shocks was the same but the voltage of the second shock had been increased, it then produced a compound action potential.

2.17 Effects of Temperature on Nerve Conduction Velocity and Refractory Period

The apparatus is arranged as described in paragraph 2.15 except that the nerve bath has a glass tube parallel and close to the nerve and a thermometer or thermistor. By passing a current of water through the tube the temperature of the bath can be altered. Measure the temperature of the bath and the conduction velocity of the fastest fibres over a temperature range 5° C to 35° C. Produce a graph relating nerve temperature to conduction velocity. Measure the slope of the graph in metres per degree C increase in temperature in the temperature ranges 5–10, 10–20, 20–30 and 30–35° C. Measure the refractory period, as described in paragraph 2.16, at different temperatures. At the end of the experiment raise the temperature above 35° C: at what point does conduction fail?

2.18 The Isolated Frog Rectus Abdominis Preparation

Stun and pith a frog. Lay it ventral side uppermost on a cork board. Lift up and remove the skin over the abdomen. The position of the rectus abdominis muscle is shown in Figure 2.15. Note that there are two muscles, one on each side of the mid-line, so two preparations can be made from one frog. With a curved needle attach threads to the ends of the muscle while it is in place. Tie a short loop on the lower thread so that it can be fitted over the glass anchoring tube which also serves as an aerator in the organ bath. The upper thread should be three or four inches long. Cut out the muscle with threads attached.

The organ bath is 10 ml. in capacity and may be made from the barrel of a 10 ml. plastic syringe by fitting a second tube to the base. The apparatus is shown in Figure 2.16. The muscle should work against a tension of about 1 g., so attach a 1 g. weight to the lever at the point where you will later fix the thread from the upper end of the muscle and add plasticine to the lever on the far side of the fulcrum until it balances. There is a further 1 g. weight attached to the pointer side of the lever by a thread; it is used to increase the load to 2 g. and stretch the muscle between contractions. This weight is rested on some convenient part of the assembly when the extra tension is not required.

Fill the bath with Frog Ringer, attach the two ends of the muscle, start the aerator to give a continuous stream of bubbles (about ten a second). Apply the stretcher weight for a few minutes.

The muscle responds to dilute solutions of acetylcholine with a slow contracture. For comparative work a standard procedure should be followed. The one recommended below has a five minute cycle, so the minute graduations of an ordinary clock set up on the bench by the bath can be used to follow the cycle.

Time	*Instruction*
0 min	Take off stretcher weight. Start kymograph.
2 min	Inject drug into organ bath.
3 min	Stop kymograph, put on stretcher weight, drain bath and refill with Ringer.
5 min	End of cycle.

Kymograph speed 2 mm per min.

Read section 1.14 on Drug Concentrations.

The stock solution of acetylcholine contains 1 mg. per ml. (10^{-3}). This solution is in distilled water with a little acid (pH 4). Take 1 ml. of the stock solution and dilute to 20 ml. with distilled water. 1 ml. of this solution thus contains 0·05 mg., the concentration is 5×10^{-5}. Add 0·1 ml. (containing 5 μg) to the bath with a syringe (1 ml.) and record the response. 5 μg in 10 ml. is a concentration of 5×10^{-7}. Go through a few cycles with varying concentrations, e.g. 1×10^{-6}, 2×10^{-6}, 2×10^{-7} to get an idea of the sensitivity of the preparation. Note that above a certain concentration no further increase in contraction is obtained. This is the maximal contraction. Over the range of submaximal contractions the response is approximately logarithmic, i.e. to double a response you must double the concentration. Plot a graph between the height of the

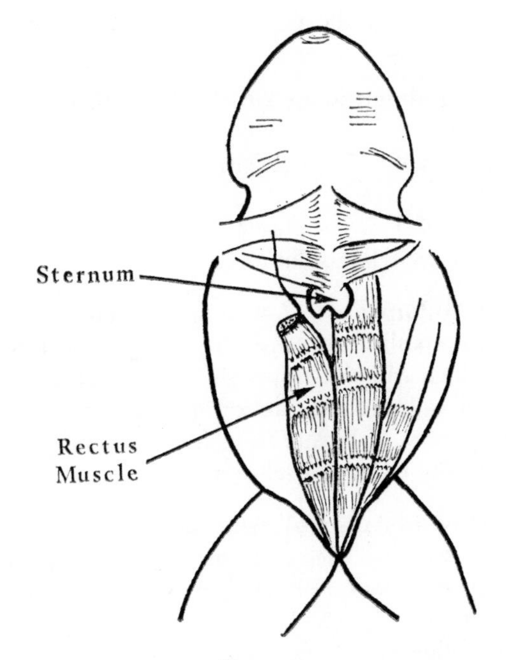

FIG. 2.15

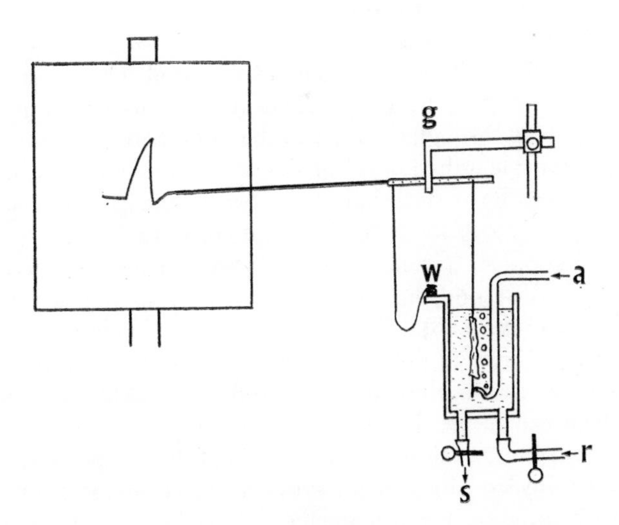

FIG. 2.16

The gimbal lever (g) is balanced so as to exert a tension of 1g on the rectus muscle. Weight (w), when applied, adds another 1g tension. Air is bubbled through the Ringer's solution via tube (a). Tube (r) is connected to a reservoir bottle of Ringer's solution and tube (s) leads to the sink.

contraction and the logarithm of the dose. This is the dose response curve of the preparation.

2.19 The Effect of (+)—Tubocurarine on the Sensitivity of the Rectus Muscle to Acetylcholine

Proceed as in 2.18 to establish the sensitivity of the preparation. Find a dose of acetylcholine which will give a response about $\frac{3}{4}$ maximal. Fill the bath with Ringer's containing (+)—tubocurarine 1×10^{-6}, leave 2 min. Run a few 5-min cycles to follow change of sensitivity to acetylcholine solution.

2.20 Use of the Rectus Abdominis to Assay an Acetylcholine Solution

Proceed as in 2.18 to establish the sensitivity of the preparation. Find the concentration of acetylcholine which will give responses about $\frac{1}{3}$ maximal and $\frac{2}{3}$ maximal. Now test the unknown solution and find the dilution necessary to obtain a response between the $\frac{1}{3}$ and $\frac{2}{3}$ maximal response. Calculate from dose response curve and your dilutions what the concentration of the unknown solution is.

2.21 To Show the Effect of an Anticholinesterase

Proceed as in 2.18 to establish the sensitivity of the preparation. Find the dose of acetylcholine which will give a response about $\frac{1}{10}$ maximal. Fill the bath with neostigmine in Frog Ringer concentration 5×10^{-8}. Leave 5 min. Run a few 5-min test cycles with acetylcholine solution to follow the change of sensitivity of the rectus muscle.

2.22 The Rat Diaphragm Preparation

Apparatus. The preparation is set up in an isolated-organ bath, a uterus-bath of 50 ml. capacity is excellent. The bath[33] should be fitted with a water-jacket and the temperature maintained at 30° C by an electric heater. A mixture of 95 per cent. O_2 and 5 per cent. CO_2 is bubbled through the fluid in the inner bath by means of a piece of fine polythene tubing. A short length of fluted glass rod inserted into the end of the polythene tubing assists the formation of fine bubbles. Alternatively, use a sintered glass gas distribution tube (see Fig. 2.17 for general plan of apparatus).

The diaphragm support[1,33] consists of an **L**-shaped piece of perspex to which is clamped at right angles a second, smaller, **L**-shaped perspex rod carrying the electrodes for nerve stimulation (Fig. 2.18). When the clamp is loosened the nerve-electrode assembly may be moved up and down the diaphragm support according to the length of nerve available. The nerve-stimulating electrodes themselves consist of two parallel platinum wires which run across the bottom of a groove cut in the short limb of the **L**, and are connected to two terminals on the upper end of the perspex rod. The dimensions of the electrode assembly are: diaphragm support, 11 cm long × 6 mm thick; foot of diaphragm support 2 cm long, 3 mm diameter; electrode support, 11 cm long × 6 mm thick; foot of electrode support 2 cm long with groove cut 3 mm from end. Electrode wires in groove, spaced about 3 mm apart.

The short horizontal limb of the diaphragm support (to which the costal margin of the preparation is tied) carries a platinum wire which acts as one electrode for direct

stimulation of the muscle. This wire is connected to a terminal near the top of the diaphragm support. The tendon of the preparation is tied to a platinum wire fixed to the short arm of the light isotonic heart lever which records the contractions (Fig. 2.17) This wire transmits the pull of the muscle and also acts as the second electrode for direct stimulation, the current passing along the diaphragm between ribs and tendon. To ensure good electrical contact a fine flexible copper wire is soldered to the platinum wire just below the lever and leads to a terminal mounted on the lever support.

Method of Making the Preparation

The rat is killed by a sharp blow on the head and is bled out by cutting the vessels in the neck; alternatively, anaesthetize with trichloroethylene vapour and bleed out. The abdomen is opened and the inferior vena cava cut through at the point where it enters the diaphragm. The skin is then reflected from both sides of the thorax and the muscles overlying the ribs are cut away. A small incision is made through the diaphragm (from the abdominal side) at its attachment to the ribs, close to one side of the sternum. One blade of a strong scissors is inserted into the incision and all the ribs are cut through close to the sternum. This procedure is repeated on the other side of the sternum. The sternum can now be carefully dissected from the mediastinal tissue and removed. A straight cut is made across the ribs on the right side parallel to the diaphragmatic attachment, starting between the fifth and sixth ribs. All the ribs above this cut are then removed. A second transverse cut is made along a curved line just below the attachment of the diaphragm to the chest wall, starting between the second and third false ribs. This leaves the diaphragm attached to a segment of the chest wall containing portions of four ribs.

The right phrenic nerve should now be visible entering the diaphragm and running upwards on the mediastinum. Gently pull the nerve medially and the lobes of the right lung laterally and cut off the lobes of the lung separately near the hilum.

Wash out the thoracic cage with Krebs' solution (or Tyrode's) and carefully dissect the right phrenic nerve from the mediastinal tissue, beginning close to the diaphragm and working up towards the neck. When the nerve has been freed as high as possible tie a ligature round it and cut above the ligature. A length of about 2·5 cm is usually obtained.

Next insert two ligatures through the costal segment which is attached to the diaphragm. The ligatures should be about 1·5 to 2 cm apart with the point of entry of the phrenic nerve lying midway between them. Pull gently on these ligatures so as to stretch the diaphragm and observe the direction in which the muscle fibres run. Make two cuts, parallel to the muscle fibres and commencing just outside the costal ligatures, through the ribs and diaphragm to converge on the central tendon. Tie a ligature round the narrow portion of tendon at the apex of the fan-shaped section thus cut out, and sever the tendon distal to the ligature. Transfer the nerve-muscle preparation to a watch-glass containing Krebs' or Tyrode's solution at room temperature. A similar preparation may, of course, be made from the left side, though it is not usually advisable to attempt to obtain two preparations from the same animal. The right side preparation

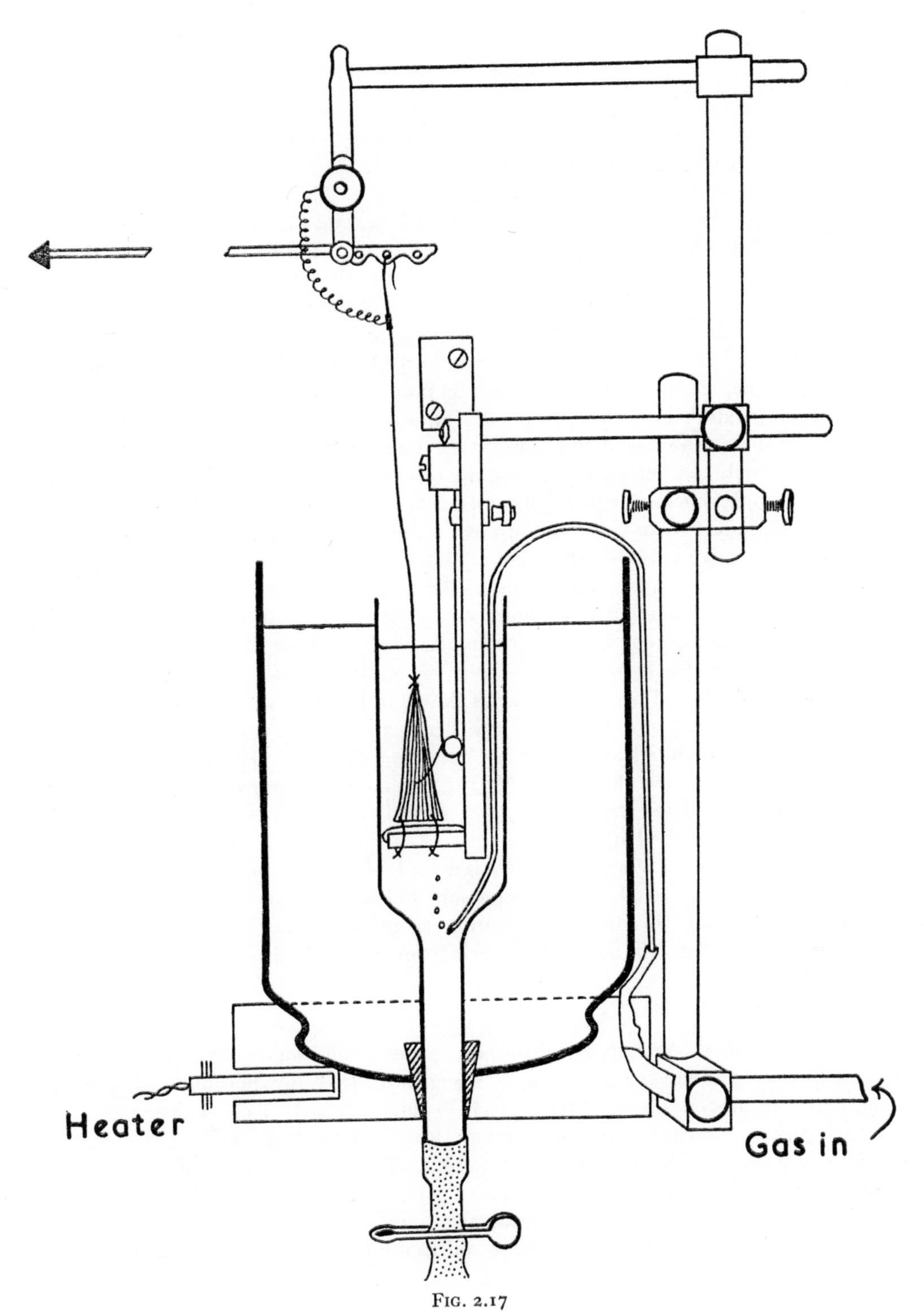

FIG. 2.17
Bath for the rat diaphragm preparation.

usually gives better contractions, though the nerve is slightly more difficult to dissect than that of the left side which is not so closely bound to the mediastinum.

When the diaphragm preparation has been made the recording lever and electrodes are raised as one unit and are swivelled clear of the bath. The tendon is tied to the hooked end of the wire which hangs from the recording lever and the two ligatures which have been passed through the costal margin are tied firmly to the horizontal arm of the diaphragm support. The phrenic nerve is then lifted into the groove in the electrode holder so as to lie across the two electrode wires. The end of the nerve is clamped between the electrode holder and the diaphragm support to hold it firmly in place. The assembly is then swung back and lowered into the fluid in the bath, so that the whole preparation, including the nerve-stimulating electrodes, is immersed.

The preparation is most conveniently stimulated by a square-wave generator in which both the strength and duration of the pulses can be controlled. The rate of stimulation should not exceed 12 per minute, otherwise fatigue may occur. For indirect stimulation single shocks of 1 to 3 volts and 0·2 msec duration are usually adequate. For direct stimulation 25 to 50 volts and 2 msec duration give adequate responses.

In an alternative assembly the diaphragm support is a glass rod with its lower end bent to fit the curve of the costal margin of the preparation which is tied to the rod by two ligatures. Glass beads may be fused on to the rod to hold the ligatures in place. The thread from the tendon is attached directly to the isotonic recording lever.

The electrodes for stimulating the phrenic nerve consist of two platinum discs mounted on an **L**-shaped perspex handle as shown in Figure 2.19. The dimensions are: platinum discs, 1 mm diam., 0·5 mm thick. Perspex handle 10 cm long, $5 \times 2{\cdot}5$ mm in section. Foot, 10 mm long, 5 mm high, 2·5 mm wide. A hole is drilled through the discs and the foot of the perspex handle so that the phrenic nerve may be pulled through by means of a ligature. The electrode assembly is fixed vertically so that it just touches the surface of the fluid in the bath.

With either method it is advisable to wait for half an hour after the preparation has been set up in the bath to allow the contractions to become steady before beginning any experiments. During this time the solution should be changed at least twice as it becomes contaminated with blood and debris from the newly dissected preparation.

Solutions Required

The preparation contracts fairly well in Tyrode's solution, preferably with double glucose (for composition see p. 13) but the height of the contraction diminishes after a time. Better results are obtained from Krebs' solution (p.13). A convenient method of making up stock Krebs' solution is as follows: to 15 litres of glass-distilled water add 108 g. NaCl, 32·76 g. $NaHCO_3$ and 5·52 g. KCl. Then add 120 ml. of a 3·82 per cent. stock solution of $MgSO_4$ and 120 ml. of a 2·11 per cent. stock solution of KH_2PO_4. This combined stock solution keeps for a considerable time. For each experiment take 500 ml. of the stock solution and add 1·0 g. glucose and 10 ml. of a 1·22 per cent. stock solution of $CaCl_2$ just before use.

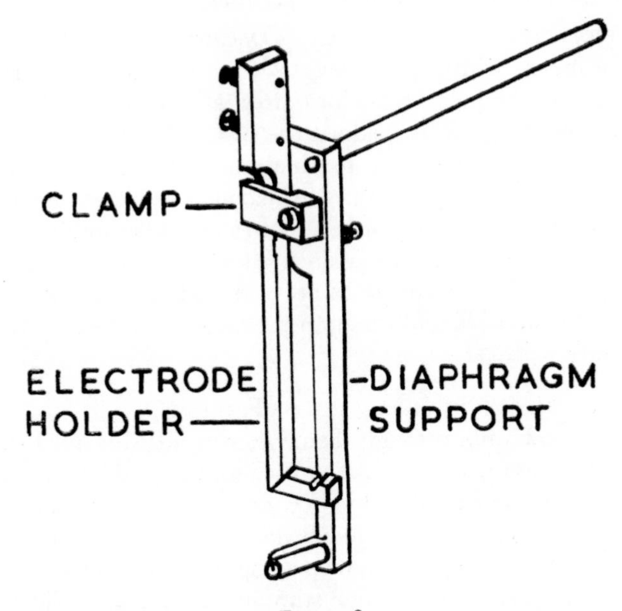

FIG. 2.18
Electrode carrier for stimulation of the phrenic nerve.

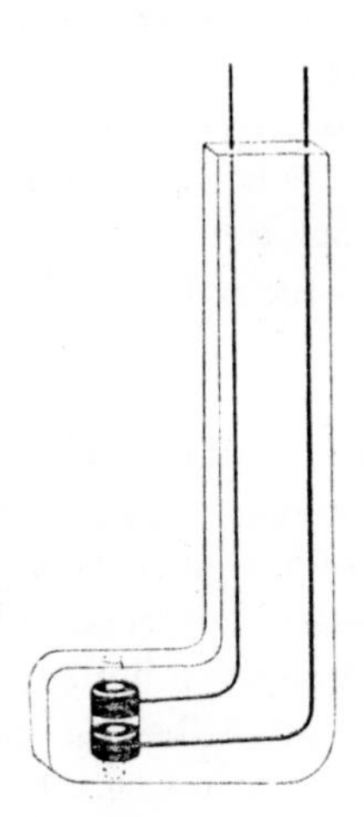

FIG. 2.19
Electrodes for stimulating the phrenic nerve.

Experiments with the Phrenic-Diaphragm Preparation

(1) Effect of Curare on Neuromuscular Transmission

Stimulate the nerve with a pulse length 0·2 msec, adjusting its strength until a maximal contraction is obtained. Switch to direct stimulation of the muscle and, using a longer duration stimulus (2 msec), adjust the strength until a comparable contraction is obtained. (It is almost impossible to obtain a maximal contraction of the muscle stimulated directly under fluid). Return to maximal nerve stimulation (0·2 msec) as before and continue to stimulate the nerve about 12 times per minute. Then add 100 μg. (+)−tubocurarine to the 50 ml. bath (a stock solution of curare containing 100 μg. per ml. is convenient). When the response to nerve stimulation has failed, change to direct stimulation, using the strength and duration previously employed. Wash several times to remove the curare.

(2) Effects of Duration of Stimulus and of Eserine

Stimulate the preparation through the nerve with just maximal shocks of 0·2 msec duration. Then, keeping the current strength constant, gradually increase the duration of the stimulus. When the stimulus duration reaches a certain value (usually between 1 and 1·5 msec), the contractions suddenly increase. This is due to summation of contractions, because two volleys are now passing down the nerve with each stimulus. Return to the just maximal short duration stimulus (0·2 msec). Add 5 μg. of eserine to the 50 ml. bath and continue the stimulation. After about 10 minutes the contractions will again increase, this time due to multiple contractions of the muscle fibres in response to a single nerve volley. Why does eserine cause this effect?

(3) Effect of Potassium Chloride

After removing the eserine by repeated washing, stimulate indirectly with just maximal shocks of 0·2 msec duration. Add 25 mg. KCl to the 50 ml. bath. The contractions first increase and then decrease until the muscle fails to respond to nerve stimulation. (If this effect is not obtained add more KCl as preparations vary in sensitivity). Wash with fresh Krebs' solution and the contractions will return to normal. Examine the effects of KCl on a curarized preparation stimulated directly.

2.23 Human Muscle Twitch

The human ulnar nerve and flexor minimi digiti proprius make a 'preparation' very like the frog muscle-nerve preparation. Prepare the apparatus as for the frog muscle experiments and follow the technique given there with the following modifications. A stimulator which will give an output up to 90 volts is necessary. See the section on stimulators in Chapter 1. Clamp the special human muscle lever, a robust version of the frog muscle lever, on the pillar instead of the frog bath. Clamp the base of the pillar to keep everything as rigid as possible. Tie a weight (50 to 100 g.) to the lever. Attach one end of a long piece of cotton thread to the lever at the hole farthest from the fulcrum, carry it over the pulley, then pass the subject's right little finger through a loop in the other end of the cotton. The bare forearm and wrist are supported on a wooden platform

carrying a metal plate covered with a cloth soaked in saline. (The concentration of salt is immaterial in this case as the fluid is acting merely as a conductor.) The plate is connected to the positive stimulator terminal and forms an indifferent electrode. The arm should be held in a position midway between supination and pronation. The experimenter must hold down the subject's wrist with his right hand to prevent the action of the flexor carpi ulnaris. A small electrode covered with chamois leather soaked in saline is connected to the other stimulator terminal. It is placed by the experimenter over the ulnar nerve as it lies in the groove on the posterior surface of the medial epicondyle of the humerus. The subject holds this electrode in position with the left hand. The experimenter controls the drum and the stimulator output. When a record is obtained mark the moment of stimulation as before. Record a time-trace.

2.24 Electrical Stimulation of Human Muscle and Nerve

Electrical excitation of human muscle and nerve is employed clinically to diagnose certain neuromuscular disorders (especially nerve injury) and to maintain contractility in temporarily paralysed muscle as in poliomyelitis. Read section on Stimulators in Chapter 1 and examine the instruments and stimulation electrodes provided.

(1) *Determination of Threshold Stimuli*

For this purpose it is not necessary to record the contraction of the muscle as in 2.23. Much information can be got by watching and feeling for the muscle tightening and the tendon moving when the stimulus is applied. Lay the forearm, comfortably relaxed, on the bench with the flexor surface upwards. Connect the plate in the jar of saline to the positive terminal of the stimulator or source of direct current, and the active electrode to the negative terminal. Set strength at 12 volts and use long pulse length. If a D.C. source is being used, a push button switch can be used to control the length of the stimulus manually. If no contraction ensues, do not at first increase the voltage, but explore the surface of the forearm with the electrode, applying one or two shocks at each position. The most effective position for provoking a muscle's contraction is called its 'motor point'; it usually lies over the point of entry of the nerve into the muscle belly and is often quite a sharply localized point. The motor points for the finger flexor muscles lie about midway between wrist and elbow, well over on the ulnar side. If no response is produced, increase the voltage in steps until a contraction becomes evident, and then identify (and mark with ink) the motor points of various muscles. Thereafter, decreasing the voltage if need be, determine the threshold voltage for minimal detectable contraction.

Notice that a twitch occurs when the current is 'made', and perhaps at 'break' as well, but if the switch is held on, no continuous or tetanic response is produced. As a rule the voltage required to produce a 'break' response is considerably higher than that needed to give a 'make' contraction. It may not be possible to obtain a figure for the threshold value of 'break' shocks in certain subjects, for the voltage required may be so high that the preliminary 'make', which is unavoidable, causes an uncomfortably violent contraction.

Now apply stimuli of short pulse length (0·2 msec) repetitively. Apply the active electrode to one of the motor spots previously located, set pulse rate to 25 and adjust its strength to give a response. Notice that the contraction is tetanic in nature, and that its strength can be varied by changing the voltage of the stimulus. Make further explorations of the forearm for motor points.

(2) *Stimulation of Sensory Nerves*

During the process of identifying the muscle 'motor points' certain regions of the skin will have been found to give rise to tingling or pricking sensations. Attempt to locate one or other of the cutaneous sensory nerves in the following way: bare the upper arm and moisten it with saline. Use the edge of the disk electrode to get as accurate a localization as possible and adjust the strength to give an easily preceptible amount of tingling when the edge is lightly pressed on the skin. Move the electrode slowly across the surface of the arm. For the most part you should feel local pricking and tingling, but sooner or later the sensation will change to a fluttering that is referred down the forearm, often as far as the wrist. If the electrode is made to explore the arm at different levels there is no difficulty in tracing and marking the position of the cutaneous nerve, often for several inches. Note that the sensation, though of an abnormal type, is referred to the peripheral distribution of the nerve. Also note that while the nerve is being stimulated, tactile sense over its whole distribution is much diminished. Why is this so?

2.25 Conduction Velocity in Human Nerve

The ulnar nerve is used for this experiment. First revise your knowledge of the course taken by the nerve between shoulder and fingers. The plan of the experiment is to stimulate the nerve, first above the elbow and then near the wrist, and measure the time taken for the motor impulses to activate muscle fibres in the hand. By subtracting the two times the conduction time between elbow and wrist can be found.

Wash the skin of the palm of the subject's hand, apply electrode jelly and silver foil electrodes (dimensions about 2 × 1 cm) separated by about 3 cm, and overlying the abductor and flexor muscles of the smallest finger. Screened wires from these electrodes are connected to the Y amplifier of the oscilloscope (differential input, AC coupling, sensitivity 1 mV/cm). Fit a large electrode, e.g. an ECG plate electrode, to the forearm and connect this to the oscilloscope earth connection. Ask the subject to flex the muscles under the pick-up electrodes and check that an electromyogram appears on the screen of the oscilloscope. Use a free-running time-base, velocity 20 msec/cm for this test.

Because a high gain amplifier is used to detect the muscle action potentials the stimulator output must be isolated from earth. Also, because the nerve is stimulated through the skin, an output of up to 60 volts may be required. The most convenient arrangement with the available stimulators is to use the 2CROS (see section on Stimulators on p.2) to trigger the oscilloscope time-base and after a delay of 1 msec the output of channel 1 (18v) is used to trigger the Devices isolated stimulator to emit a 0·5 msec pulse which is applied to the stimulating electrodes.

Apply the anode silver foil electrode to the skin over the biceps muscle. Use electrode jelly to ensure a low resistance connection. With a repetition frequency of 2 per second deliver shocks between the anode and an exploring electrode, which has an insulated handle and a moist, saline-soaked pad. Move the exploring electrode about until the point is found where a movement of the little finger can be obtained with the lowest voltage setting of the stimulator. Does this point correspond to the supposed course of the subject's ulnar nerve? Mark the point with a wax pencil.

Move your stimulating electrodes to the wrist and mark the point where the exploring electrode produces a similar movement of the finger. The stimulating voltage may be higher and firm pressure is required to find the right position.

Examine the wave form of the response on the triggered time-base, velocity 2 msec/cm. Measure the time interval between the shock artifact and the beginning of the muscle action potential in each of the two stimulating positions. Measure the distance between the two stimulating sites by laying a thread or wire along the nerve path. Calculate the velocity of the motor nerve impulses thus:

$$\frac{\text{Distance between stimulating sites (mm)}}{\text{Conduction time (msec) elbow to finger—Conduction time (msec) wrist to finger}}$$

i.e. mm/msec or m/sec.

2.26 The Movements of the Uterus *in Vitro*

These movements can be recorded exactly as described for the gut (see paragraph 9.4). The virgin guinea-pig uterus is used to compare solutions containing an unknown quantity of oxytocin with standard solutions—a biological assay.

Since the movements are very slow, use a drum speed of 0·1 mm per sec. The virgin guinea-pig uterus is quiescent at the beginning of an experiment, showing only minor variations of the base-line. When posterior pituitary extract (about 0·03 oxytocic unit) is added to the bath the uterus contracts after a short latent period. Study also the action of adrenaline and histamine on this preparation. If virgin guinea-pigs are not available, stock animals will usually give very active uteri. The amplitude of the spontaneous variations is less if a half horn instead of a whole horn is used.

2.27 The Rat Uterus Preparation

The spontaneous contractions of the rat uterus can be suppressed by the use of a modified Locke's solution called de Jalon's solution; it had a composition NaCl, 9 g.; KCl, 0·42 g.; $CaCl_2$, 0·06 g.; glucose, 0·5 g.; $NaHCO_3$, 0·5 g.; in 1 litre distilled water. Find out the concentration of acetylcholine which will give a sub-maximal contraction. Try 10 to 20 μg. in a 50 ml. bath. Show how this response is diminished by low concentrations of adrenaline (10^{-8}) added to the bath 30 seconds before the acetylcholine. This experiment shows the basis of a sensitive method of assay of adrenaline.

CHAPTER THREE

RESPIRATION

3.0 Introduction

This chapter begins with a description of the rather specialized nomenclature and symbols which are used in respiratory physiology. Then the subdivisions of the total lung capacity are considered together with an introduction to the measurement of gas volumes. Tests of lung function follow and then details are given of the methods of gas analysis which have proved useful in respiratory work.

There is some overlapping between chapters, of experiments employing measurements of pulmonary ventilation and oxygen consumption. The respiratory quotient and experiments relevant to the control of respiration are described in the latter part of this chapter, whereas estimation of metabolic rates and the metabolic cost of muscular work by respiratory techniques are considered in Chapters 4 and 10 respectively.

3.1 Nomenclature of Respiratory Physiology

General agreement has now been reached on the subdivisions of total lung capacity. These are based on the assumption that the most stable point of the respiratory cycle is the resting end-expiratory position, i.e. during quiet respiration we tend to breathe out to the same functional residual capacity each time (Fig. 3.1). The word 'volume' has been reserved for discrete fractions of the total lung capacity; capacity is used where the measurements can be divided into smaller entities, e.g. Vital Capacity includes both Inspiratory Capacity and Expiratory Reserve Volume.

LUNG VOLUMES

VC—Vital Capacity—Maximal volume that can be expired after a maximal inspiration.

IRV—Inspiratory Reserve Volume—Maximal volume which can be inspired from end-tidal inspiration.

ERV—Expiratory Reserve Volume—Maximal volume which can be expired from the resting end-expiratory level.

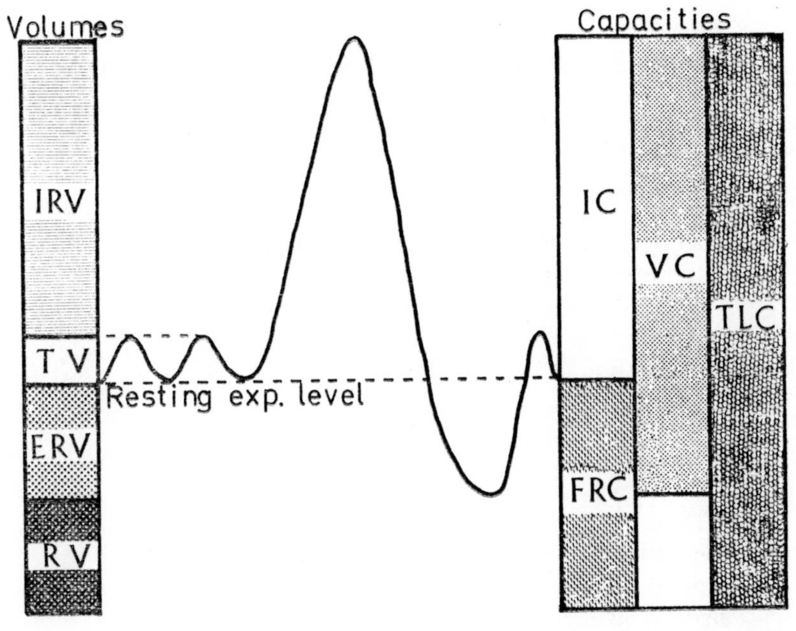

Fig. 3.1

IC—Inspiratory Capacity—Maximal volume which can be inspired from the resting end-expiratory level.

FRC—Functional Residual Capacity—Volume of gas in the lungs at the resting end-expiratory level.

RV—Residual Volume—Volume of gas in the lungs at the end of maximal expiration.

TLC—Total Lung Capacity—Volume of gas in the lungs at the end of a maximal inspiration.

This branch of physiology has been further systematized by the development of a shorthand notation. This consists of a list of abbreviations and a convention for their use; a list is shown in Table 3.1.

The Primary Variable, denoted by the first symbol, is the parameter to be measured; it is represented by a large letter, usually a capital. It is followed by the Secondary Variable, which denotes where the primary variable was applied. This is given as a small capital for quantities in the gas phase and a lower case letter for the blood phase and these symbols are written on the same line as the primary variable. The particular gas measured is printed next, as a subscript, using the normal chemical symbol. Finally a dot above a primary symbol indicates a time derivative and a short dash indicates a mean value.

TABLE 3.1

SYMBOLS USED IN RESPIRATORY PHYSIOLOGY

Primary Symbols

P = pressure
V = volume
F = fractional concentration in the dry gas phase
f = respiratory frequency
Q = volume of blood
C = concentration of gas in blood phase
S = % saturation of haemoglobin with O_2

Secondary Symbols

(1) Gas Phase

I = inspired air
E = expired air
A = alveolar air
T = tidal air
D = dead space air
B = barometric

(2) Blood Phase

a = arterial blood
v = venous blood
c = capillary blood

Symbols used in Respiratory Physiology based on Pappenheimer, J. *et al.* (1950). *Fed. Proc.*, **9**, 602-605, Gandevia, B. & Hugh-Jones, P. P. (1957) *Thorax*, **12**, 280-293,

Thus,

$\overline{P_B}$ = Mean Barometric Pressure

$\dot{V}_A$ = Alveolar ventilation per minute

Pa_{CO_2} = Partial pressure of CO_2 in arterial blood

F_{IO_2} = Fractional concentration of O_2 in inspired air

P_B = Barometric Pressure

3.2 Spirometry

Many measurements of lung volumes and ventilation are obtained with the use of a spirometer. This is simply a counterpoised gasholder. A writing-point fixed to the counterweight is made to write on a variable-speed kymograph drum. The ideal spirometer offers so little resistance to the movement of air to and from the subject's lungs that its presence is imperceptible and the subject's respiratory movements are unaffected by the apparatus. This ideal is difficult to achieve but it is approached by the use of wide-bore airways, light-weight gas bells, and low-inertia recording devices. Soda-lime absorbers and one-way valves introduce airway resistance and are best dispensed with if possible. They are necessary, of course, where the spirometer is used to measure oxygen consumption by the closed circuit method (para. 4.2) and here some lack of fidelity in the record can be tolerated. Some improvement is obtained by using a pump to circulate the gas through the carbon dioxide absorbent. The effects of airway resistance are well illustrated by Campbell *et al.* (1961). *Clin. Sci.* **21,** 311.

It is possible to avoid the use of a soda-lime absorber in two ways. Firstly, use the spirometer only to collect and measure the expired air, as in paragraph 3.6 or in FEV tests, or secondly, limit the duration of rebreathing from a large capacity spirometer so that carbon dioxide accumulation is unimportant. When measuring MVV as in paragraph 3.10 an accumulation of carbon dioxide is actually helpful.

For tests of short duration, air is used to fill the spirometer. Pure oxygen depresses ventilation even in the normal subject and in abnormal conditions where there has been some persistant anoxic drive it may produce marked depression. In longer lasting tests where oxygen consumption is being studied rather than respiratory rhythm and minute ventilation, pure oxygen is used to fill the spirometer.

Some further points of spirometric technique must also be considered. Consecutive determinations of any lung volume must be made with the subject in the same position. The effect of posture is illustrated by its effect on resting end-expiratory position in paragraph 3.5. Before starting a recording, the bell should be raised and lowered with the ink-writer against the record paper. This provides a vertical axis to assist measurements on the paper. Make sure the spirometer is level and that the bell does not touch the sides.

Note that gas cylinders are provided with two valves, a main cock attached to the cylinder itself and a secondary valve which is screwed into the cylinder orifice. The main cock is used to seal the cylinder when it is not in use. The secondary valve may automatically regulate the output pressure to a few pounds per sq. in. giving variable volume flow at a constant pressure or it may be a simple needle valve which must be adjusted manually to give the correct output pressure, and volume flow depends on the pressure selected.

Several types of mask or mouthpiece are used to connect the subject to the spirometer. Short-lasting tests which involve high airflow rates such as FEV are best performed with a mask pressed against the face to include mouth and nose. In other tests where a small leak will cause great errors as in closed-circuit oxygen consumption measurements, a rubber flanged tube is fitted between the lips and teeth and gripped by the subject. For many tests a 3 cm diameter glass tube gripped by the lips is suitable.

3·3 Gas Laws: Volume Conversions

The respiratory gases obey the fundamental physical laws governing the behaviour of gases. The most important of these are:

1. *Boyle's Law:* The volume occupied by a quantity of gas is inversely proportional to its absolute pressure, if its temperature remains constant.

i.e. $P_1V_1 = P_2V_2$, where $T_1 = T_2$

2. *Charles' Law:* The volume occupied by a quantity of gas is directly proportional to its absolute temperature, if the pressure of the gas remains constant.

i.e. $V_1/T_1 = V_2/T_2$, where $P_1 = P_2$

1 and 2 can be combined, $\frac{P_1V_1}{T_1} = \frac{P_2V_2}{T_2}$

Absolute temperature is recorded on the Kelvin scale (° K). Zero on the scale is equivalent to −273° C and is the temperature at which any body would be incapable of releasing further thermal energy. Absolute pressure is the actual pressure at a point in a fluid (gas or liquid).

3. *Dalton's Law of Partial Pressure:* If several gases are placed in the same container, the total pressure exerted is the sum of the partial pressures which each gas would exert if it alone occupied the container.

i.e. $P_{Total} = P_1 + P_2 + P_3 \ldots + P_n$

Alternatively, if n_A moles of gas A were alone in a volume V at temperature T, the pressure would be

$P_A = n_A . RT/V$ (where R = volume occupied by 1 mole of any gas at 0° C and 1 Atmos. pressure)

and similarly $P_B = n_B . RT/V$

$\therefore P_A/P_B = n_A/n_B$

e.g. in dry air 20·93 per cent. of the molecules are O_2

$Po_2/P_{Total} = 20{\cdot}93/100$

∴ Partial pressure of O_2 at a pressure of 760 mm Hg

$= \frac{760 \times 20{\cdot}93}{100} = 159{\cdot}1$ mm Hg

4. At a gas/liquid interface created by exposing a liquid to a vacuum, the number of molecules with sufficient kinetic energy to evaporate (per second per unit area of liquid surface) is proportional to the temperature of the liquid. The number of molecules which return from the gaseous to the liquid phase is that number which strike unit area of the surface in one second, and at a given temperature this is proportional to the vapour pressure.

Since rate of evaporation = rate of condensation at equilibrium, it follows that the vapour pressure is determined by the temperature.

When an interface exists between a liquid and a permanent gas (or gases) e.g. air, such an equilibrium state is reached. Then, by Dalton's law,

P_{Total} = Partial pressure exerted by each gas + partial pressure exerted by the fluid in the gaseous phase.

In physiological practice, this means the pressure exerted by water vapour. Air is saturated with water vapour when the partial pressure exerted by the vapour is maximal for a given temperature.

5. *Solubility of Gases in Liquids: Henry's Law*

The quantity of gas physically dissolved in a liquid at constant temperature is directly proportional to the partial pressure of the gas in the gas phase. The solubility coefficient for a gas in a particular liquid is the amount of the gas which will dissolve in unit volume of liquid per unit in partial pressure, e.g. the solubility coefficient of CO_2 in water is 0·0334, i.e. if P = partial pressure of CO_2 in mm Hg, 0·0334P m-moles of CO_2 will dissolve per kg. of H_2O at 37° C.

The partial pressure of a gas in a liquid is called its tension and is not measured directly but is obtained by measuring the partial pressure of the gas in an equilibrated gas phase. Note, too, that the partial pressure of a gas in a liquid is related only to the amount in physical solution; the content of gas in a liquid may be considerably higher, e.g. CO_2 in blood, if gas is also held in chemical solution. Such chemically combined gas does not contribute to the partial pressure.

Applying these concepts, respiratory volumes can be specified in terms of the pressure, temperature and water vapour saturation and any two volumes can be equated by reducing them to similar conditions with respect to these variables. Thus volumes may be expressed as:

(1) Ambient Temperature and Pressure, Saturated with water vapour—ATPS, e.g. the volume change in a spirometer (para. 3.2).

(2) Body Temperature and Pressure, Saturated with water vapour—BTPS, e.g. the volume of one of the subdivisions of lung.

(3) Standard Temperature and Pressure, Dry—STPD. This can be used as a reference scale to which (1) and (2) are converted to facilitate equation of volumes. Standard (previously Normal) Temperature is 273° K (0° C) and Standard Pressure is 760 mm Hg.

The principal conversions used are:

(1) ATPS to BTPS

$$V_{BTPS} = V_{ATPS} \times \frac{P_B - P_{H_2O}}{P_B - 47} \times \frac{310}{273 + t^\circ C}$$

Since small deviations from standard barometric pressure produce little change in the final result, this can be simplified to a correction for temperature only and Table 3.2 gives a list of conversion factors for the normal range of temperatures encountered.

TABLE 3.2

Factor to convert Volume to 37° C Sat.	When Gas Temperature Saturated at ° C is	Saturated Vapour Pressure (mm Hg)
1·112	18	15·5
1·107	19	16·5
1·102	20	17·5
1·096	21	18·7
1·091	22	19·8
1·085	23	21·1
1·080	24	22·4
1·075	25	23·8
1·068	26	25·2
1·063	27	26·7
1·057	28	28·3
1·051	29	30·0
1·045	30	31·8
1·039	31	33·7
1·032	32	35·7
1·026	33	37·7
1·020	34	39·9
1·014	35	42·2
1·007	36	44·6
1·000	37	47·1

(2) ATPS to STPD

$$V_{STPD} = V_{ATPS} \times \frac{P_B - P_{H_2O}}{760} \times \frac{273}{(273 + t^\circ C)}$$

The nomogram, Figure 3.2, gives a correcting factor over the normal ranges of pressure and temperature employed.

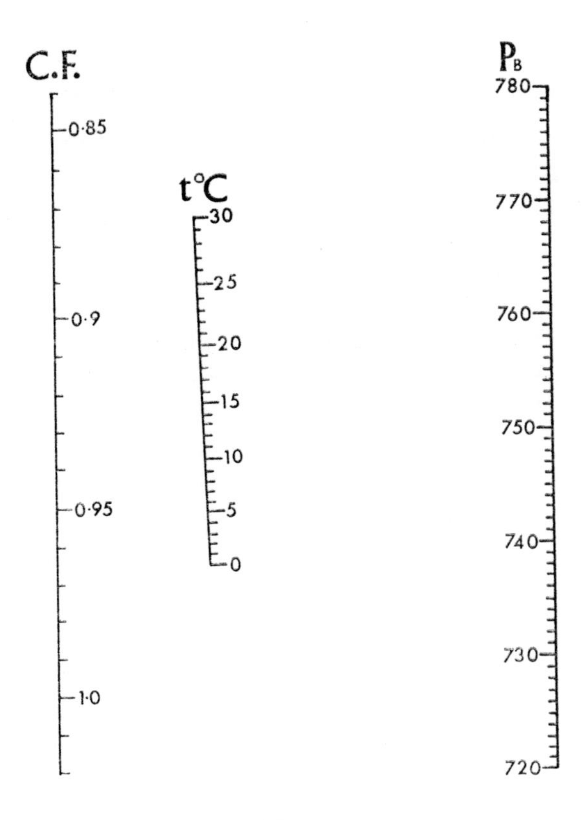

FIG. 3.2

Nomogram to obtain the correcting factor C.F. by which the volume of a gas at any pressure between 720 and 780 mm Hg and saturated with water vapour at any temperature between 0° and 30° C can be reduced to STPD. (Weir, J. B. de V. (1949). *J. Physiol.*, **109**, 1.)

3.4 'Normal' Ranges: Prediction Formulae and Nomograms

Most ventilatory measurements are made quite easily but a recurring problem is the decision as to whether the values obtained for a particular subject lie within the limits of normality. To facilitate such decisions, formulae have been derived to express correlations between particular respiratory quantities and factors such as sex, height, weight or chest size. Examples of such prediction formulae are given in Table 3.3.

TABLE 3.3

PREDICTION FORMULAE

Males:	VC litres	= ·052 Height cm − ·022 Age − 3·60
	FEV$_1$ litres	= ·037 Height cm − ·028 Age − 1·59
	MVV litres/min	= 1·34 Height cm − 1·26 Age − 21·4
	PFR litres/min	= [3·95 − (·015 Age)] × Height cm
Females:	VC litres	= ·041 Height cm − ·018 Age − 2·68
	FEV$_1$ litres	= ·028 Height cm − ·021 Age − ·86
	MVV litres/min	= (71·3 − ·47 Age) × Surface Area m^2
	PFR litres/min	= [2·93 − (·007 Age)] × Height cm

Determine your height without shoes and weight without clothes (clothed weight less 10 lb. for males and 6 lb. for females). Substitute these values in the appropriate prediction formulae and estimate the values you hope to record in later experiments.

The relation between surface area and your height and weight is given by the nomogram Figure 4.2. Fill in your measurements and predictions in Figure 3.3. below.

Age	Sex	Height	Weight	Surface Area

Predicted value	VC	FEV$_1$	MVV	PFR

FIG. 3.3

3.5 The Effect of Posture on Functional Residual Capacity

Set up the Benedict-Roth closed circuit spirometer as described in paragraph 4.2. Record for four minutes with the subject lying supine on the couch, then instruct the subject, still with his mounthpiece in position, to stand up gently with as little disturbance as possible to the spirometer and airlines. The record at this point should show a rise due to the increase in functional residual volume brought about by the descent of the abdominal contents and diaphragm under the influence of gravity. After a few minutes in the standing position the subject carefully resumes the supine position on the couch. Mark the record at the appropriate points and discuss your results.

3.6 Tidal Volume

This is the volume of air passing into or out of the lungs at each respiration. It is notoriously difficult to measure the normal resting tidal volume accurately because, apart from the considerations of paragraph 3.2, as soon as someone thinks of his breathing pattern (an inevitable accompaniment of use of nose-clip, mouthpiece and spirometer) he tends to hyperventilate. This effect can be minimized by repeating the experimental procedure several times.

Fit the subject with a nose-clip and a rubber mouthpiece connected to a directional valve of the type illustrated in Figure 3.4. He is allowed to inspire from the atmosphere and the expiratory end is connected to a spirometer. Alternatively, use a short wide glass tube as a mouthpiece and allow the subject to rebreathe from a large capacity air-filled spirometer. In each case, use a low-resistance recording spirometer without a CO_2 absorber. Instruct the subject to breathe quietly and record several breaths. Take several records and, if the rebreathing method is used flush out the spirometer between each. Continue till consistent records are obtained. Take the average of several breaths and express the resting tidal volume in litres. Count the respiratory rate.

Multiply the tidal volume by the number of breaths per minute. This gives the ventilation for one minute, normally called the Minute Volume (para. 3.9).

3.7 Vital Capacity

1. Sterilize and rinse a mouthpiece, and fit it to the tubing of a spirometer. See that the bell is at its lowest position i.e. the spirometer is empty. Expire and inspire as fully as possible from the atmosphere, then while holding the nose, expire as deeply as possible, but without undue haste, into the spirometer. The volume measured in this way is the Vital Capacity.

2. It is useful to measure the vital capacity in another way, along with the other lung volumes, so that their inter-relationship is better appreciated. The bell is filled with room air and positioned so that the pen is recording in the middle part of the paper. Set the drum speed at approximately 30 mm per second. The subject inserts the mouthpiece having first put on a nose-clip. He is instructed to breathe normally eight or nine times into the spirometer and then inhale until the maximal inspiratory position is recorded. Then he exhales smoothly and completely until exhalation is complete, and he finishes with a few normal respirations. The spirometer is emptied and refilled with room air. The subject again breathes normally to and from the bell for several breaths, takes a maximal inspiration but then returns to normal breathing for a few breaths after which he breathes out maximally.

From these records the TV, IRV, ERV, IC and VC can be measured. In addition, if the ERV and the IC from the second recording are added, they should equal the value obtained for the VC in the first test. When a carefully performed two-stage VC gives values substantially greater than those obtained with the single breath method, the subject is probably suffering from obstructive airway disease.

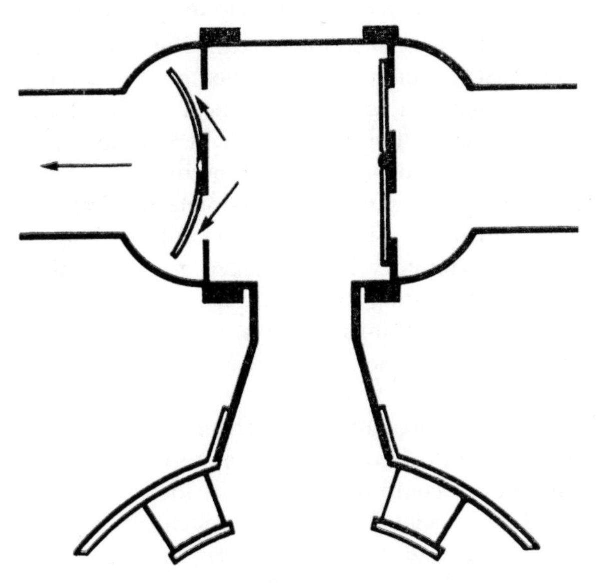

FIG. 3.4

Mouthpiece and valve assembly for respiratory work. The rubber mouthpiece is gripped between the teeth. During inspiration air is drawn through the right-hand valve and the left-hand valve is closed. During expiration, air passes out through the left-hand tube only. The moving parts of the valve are either circular rubber flaps, as shown here, or spring loaded plastic discs.

3.8 Residual Volume, Functional Residual Capacity and Total Lung Capacity

As the residual volume is that volume remaining in the lungs after a maximal exhalation, it is evident that it cannot be determined by direct spirometry. In addition, since the residual volume forms a part of both the Functional Residual Capacity and the Total Lung Capacity, these volumes too must be determined by indirect means.

The lung volume measured depends on the point at which the test is started. If collection is started precisely at the end of a complete inspiration, then the total lung capacity is the volume measured. If it is begun at full expiration, residual volume is measured. If the resting expiratory level is taken as the starting point, the volume determined is the functional residual capacity. As the end of a normal expiration is a more constant point than either full inspiration or full expiration, it is most usual to measure the functional residual capacity.

1. *Open method:* This is based on the fact that when air is breathed, gas in the lungs is 80 per cent. nitrogen. If the amount of nitrogen in the lungs at a specific position of the breathing cycle could be determined, then the total volume of gas in the lungs at that time could be calculated. All the nitrogen can be flushed out of the lungs by breathing N_2-free oxygen and collecting the expired air in a Douglas bag. The total volume of expired air and its nitrogen concentration are measured and so the total volume of nitrogen coming from the lungs is calculated.

Procedure: Two Douglas bags are required. Fit a large bore 2-way tap to each bag and then connect one tap to the inlet and one to the outlet of a 3-way valve and mouthpiece by corrugated rubber piping. Connect the oxygen cylinder to the side tube of the bag which is attached to the inlet of the mouthpiece. Flush out the whole system several times so that it is nitrogen-free. Close the tap on the inlet bag and fill the bag with oxygen. Make sure that the side tubes of both bags are closed. The subject then adjusts the nose-clip and inserts the mouthpiece. The tap of the mouthpiece is open to the atmosphere and the subject breathes to and from the atmosphere until his respiration has settled. At the end of a normal expiration, the mouthpiece is closed and both large taps on the Douglas bag are opened so that the subject inspires from the oxygen-filled bag and expires into the collecting bag. The subject should continue to breathe through the apparatus for seven minutes; the large tap on the collecting bag is then closed.

Introduce a sample of expired air from the bag into a gas analyser through the narrow side tube. After analysis of the sample the large tube of the Douglas bag is attached to a gas meter, and using gentle pressure on the bag, the volume of the expired air is measured. The volume of the expired air used in analysis may be neglected as it should be quite small.

Sample calculation:

Volume of expired air	$= 40$ litres	
Conc. of N_2 in expired air	$= 5$ per cent.	
Amount of N_2 in lungs (at resting exp. level)	$= \left(\frac{5}{100} \times 40\right)$	$= 2$ litres
But conc. of N_2 in lungs	$= 80$ per cent.	
Functional residual capacity	$= \left(2 \times \frac{100}{80}\right)$	$= 2{\cdot}5$ litres

2. *Closed Method:* With this method we make use of a closed circuit and a gas not normally present in the lungs. This method depends on the fact that we start with a known volume (the spirometer)[1] containing a known concentration of Helium, and an unknown volume (FRC) containing a known concentration of Helium (0 per cent.). After rebreathing for a short time in this closed system, the gases come into equilibrium. The new concentration of helium in the system is then measured. As theoretically no helium is lost from the system during rebreathing, the amount of helium in the lungs and in the spirometer at the beginning of the test is the same as the amount in the lungs and the spirometer at the end of the test.

Procedure: The apparatus consists of a spirometer, with a circulating pump, and a Katharometer[25] which measures the percentage of Helium in moist air. Switch on the circulating pump and the Katharometer, turning the switch below the Helium indicator to the 'TEST' position, and leave running for 10 min. Use the control 'RHEO' to bring the pointer to the red TEST mark, and then switch to 'HELIUM IN AIR' position. If there is no helium in the spirometer the pointer should indicate ZERO; if not, open the mouthpiece valve and raise and lower the bell a few times to wash out any helium which may have been left in the circuit. Always recheck the TEST position before and after taking a reading. Switch off the circulating pump.

Turn the mouthpiece valve so that the spirometer is open to the outside air, then gently lift the bell until the spirometer is half full. Close the mouthpiece valve and turn the kymograph by hand one inch to inscribe a baseline. Now add about 600 ml. of Helium and again inscribe a line on the kymograph. Switch on the circulating pump and, after equilibration, take a meter reading—C_1.

The seated subject, wearing a nose-clip, is made to breathe through the mouthpiece to and from the atmosphere. Set the kymograph moving at ·04 cm/sec and at the end of a normal expiration (resting expiratory position) the subject is switched into the circuit with the mouthpiece valve and rebreathes from the circuit. During rebreathing, oxygen is added to the circuit at approximately 300 ml./min, adjusting the flow to keep the volume at the end of expiration at a constant level. Equilibration is considered to have been achieved when the concentration does not change for 2 minutes. This will take about 5 to 10 minutes. Note the meter reading when equilibrium is reached—C_2. Before being disconnected from the spirometer, the subject is asked to give a maximal expiration, followed immediately by a maximal inspiration. This manoeuvre is repeated twice. From the tracing the Vital Capacity and Expiratory Reserve Volume can be measured, so enabling one to calculate the Residual Volume, Functional Residual Capacity or the Total Lung Capacity.

Remove the paper from the kymograph and measure in mm the distance between the base line and the helium line. In this way the volume of helium added can be obtained by multiplying the displacement distances by the spirometer calibration factor—20 ml. per mm.

C_1—first reading of meter
C_2—second reading
V_H—volume of helium added
V_S—initial volume of air in the spirometer

The initial volume of air in the spirometer—V_S—is obtained from the dilution of helium.

$$V_H = V_S \times C_1\%$$

$$= V_S \times \frac{C_1}{100}$$

$$V_S = \frac{V_H \times 100}{C_1}$$

The calculation of FRC depends on the fact that the lungs and the spirometer form one circuit and no helium is lost from it, so the amount at the beginning is the same as the amount at the end, it has only been redistributed. FRC is calculated from the following equation:

$$V_S \times \frac{C_1}{100} = (\text{FRC} + V_S) \times \frac{C_2}{100}$$

$$V_S \left(\frac{C_1}{100} - \frac{C_2}{100}\right) = \text{FRC} \times \frac{C_2}{100}$$

$$\text{FRC} = V_S \left(\frac{C_1}{100} - \frac{C_2}{100}\right) \times \frac{100}{C_2}$$

3. *Body Plethysmograph Method:* This measures the volume of gas in the thorax (V) whether this is in free communication with the airways or not. A body plethysmograph is an air-tight chamber in which pressure changes caused by respiratory movements can be measured with a sensitive manometer. When the subject breathes against a closed shutter the changes in the plethysmograph pressure reflect the thoracic volume changes ($\triangle V$), while the changes in mouth pressure measured by a second manometer represent alveolar pressure changes ($\triangle P$) as there is no actual flow of gas. The initial pressure in the lungs is atmospheric (P). From Boyle's Law $V = P. \triangle V/\triangle P$ This last term is measured as the slope of a line on an oscilloscope that receives the output of the manometers.

3.9 Pulmonary Ventilation

If pulmonary ventilation is to be measured when the subject is moving about, then use of a spirometer is no longer possible. A simple method is to collect the expired air in a Douglas bag over a timed interval and measure the volume collected by emptying the bag through a gas meter.

The bag is mounted on the subject's back, a nose-clip is applied, and by means of a mouthpiece with non-return valves the expired air is collected in the Douglas bag (Fig. 3.4).

Collect expired air over a suitable time interval (a) with subject at rest; (b) while walking briskly about the laboratory; (c) while climbing the stairs.

Express result as litres of air per minute.

There are alternatives to the Douglas bag method of measuring ventilation. A very light gas meter or Respirometer_{13} can be carried on the subject's back. This device also has an arrangement for sampling the expired air for subsequent analysis. This is described in paragraph 4.4. If the subject does not have to move about, a low-resistance gas meter such as the Crompton Parkinson CD4_{36} can be used, though care must be taken to avoid water condensation inside the meter, i.e. either interpose a water condenser or use the meter to measure inspired air. However all the methods described tend to alter the subject's normal pattern of breathing because of a cumbersome arrangement of tubes and a significant external resistance to air flow. The pneumotachograph, see paragraph 3.13, goes some way to eliminating these. Its output is related to air flow, but by integrating flow with respect to time, air volume can be measured.

3.10 Maximum Voluntary Ventilation (MVV)

This measure was formerly called the Maximum Breathing Capacity, and it is that volume of gas which can be breathed by maximum voluntary effort in one minute. In this test, the subject breathes as deeply and as rapidly as he can through a low-resistance system for 15 seconds. He is permitted to choose his own frequency and tidal volume, but in fit young people the respiratory rate should be at least 70 per minute. The maximum voluntary ventilation is a strenuous test and will give an index of the maximum breathing capacity only if the subject is fully co-operative.

1. Set up a spirometer of the low-resistance, Bernstein type, and set the recording drum at a fairly fast speed. Add air to the spirometer until the pen is recording in the mid-portion of the paper. A soda-lime canister is not used as this adds to the resistance and a slight build-up of CO_2 is advantageous. The subject adjusts the mouthpiece and nose-clip, and the mouthpiece tap is turned to allow the subject to breathe in and out of the bell. Start the recording drum and record one or two normal respirations. The subject then breathes as deeply and as rapidly as he can for 15 seconds. Switch off the motor and remove the paper. Horizontal lines are drawn which will pass through most of the inspiratory and expiratory peaks. The distance between these lines represents the average volume and this is multiplied by the number of breaths in the 15 second period. This result is in turn multiplied by 4/1000 so that the answer is expressed in litres/minute.

2. The subject takes a maximal inspiration, fits a nose-clip and then breathes as rapidly and as deeply as possible for 15 seconds into a rebreathing bag mounted within an integrating plethysmograph. In this apparatus a volume of air equal to each breath is passed through a gas meter and the pulmonary ventilation is read directly from a dial calibrated in litres.

Compare your own maximum voluntary ventilation with the ventilation achieved during fast stair-climbing and with your minute volume at rest.

The MVV is usually found to be 20 times the minute volume at rest and this relationship gives some idea of pulmonary reserve.

3.11 Forced Inspiratory and Expiratory Spirograms

Many feeble and ill subjects are unable to perform as strenuous a test as the MVV, and so there has been a tendency to replace it with the recording of a single forced inspiration, or, more commonly, a single forced expiration. Many methods of recording and measuring these spirograms have been proposed; the most commonly used measurements can be obtained with the following technique.

Empty the air from the bell of a low resistance spirometer and set the kymograph to a fast speed. Wearing a nose-clip, the subject takes as deep a breath as possible from the atmosphere, inserts the mouthpiece and blows as hard as possible into the spirometer. At the end of expiration, the subject breathes in, as fast and as fully as possible, from the spirometer. Stop the drum and remove the mouthpiece.

Note the slowing which occurs at the end of exhalation. On inspiration, the rate is maintained until the lungs are full.

One measure frequently made on the expiratory curve is the timed forced expiratory volume, e.g. FEV_1—which represents the volume of gas expired in one second from the starting time. The FEV_1 will obviously be reduced in subjects with increased airway resistance. The amount of air expired from full inspiration to full expiration is of course the Vital Capacity, in this instance as it is done as quickly as possible it is called the Forced Vital Capacity—FVC. In a normal person the FEV_1 should be at least 70 per cent. of the FVC. If a person has restrictive pulmonary disease the FEV_1 and the FVC will both be less than the predicted values, but the FEV per cent. will be normal. In a person with increased airway resistance the FEV_1 and the FEV per cent. will be less than normal but the FVC may be normal.

Many workers have multiplied the FEV_1 by various factors in order to obtain an estimate of the MVV—the so-called Indirect MVV. Most investigators now prefer to use both tests or to rely solely on the measurement of a single forced expiration.

Because the starting point of the FEV_1 may be doubtful, some authorities now use the Forced Mid-expiratory Flow as a measure of the efficiency of ventilation. This is the average rate of gas flow during the middle half-volume of the FEV (i.e. from 25 to 75 per cent. of the expired volume).

3.12 Peak Flow Rate (PFR)

Although spirometer tests have the advantage that a permanent record is obtained, their big disadvantage is that the apparatus is not very portable. A simple and portable device, the Wright Peak Flow Meter[20], is very useful in mass surveys of lung function and as a screening test in suspected obstructive airway disease.

This device makes use of a single forced expiration but differs from most other tests of this kind in that, instead of measuring the volume expired in a given time, it measures the maximum flow rate or 'peak flow'.

The subject takes a deep breath, puts the mouthpiece into his mouth and blows hard. A short sharp blast with some 'follow through' is required, but the lungs need not be emptied as in the FVC test. Return the pointer to zero after use by depressing the button beside the mouthpiece.

Take five readings and record the mean of the three highest readings.

3.13 The Pneumotachograph

This [26,28] consists of a wide bore tube through which the subject breathes but across which is arranged either a fine wire grid (Lilly type) or a series of short parallel tubes (Fleisch type). These set up a very slight resistance and the pressure-drop across them, which is proportional to the rate of air flow, can be measured by a sensitive manometer. Electrical integration of this signal gives a record of the volume of air passing the pneumotachograph. Examine the instrument and assess its suitability for measurements of peak flow rate, paragraph 3.12, tidal volume, paragraph 3.6 and vital capacity, paragraph 3.7.

3.14 Gas Distribution

Even in normal persons all parts of the lung are not uniformly ventilated. In certain lung diseases the ventilation may be very uneven.

Set up the apparatus as in experiment 3.8(1) but replace the Douglas bag used for collecting expired air with a 4-litre bag. Flush out the whole system carefully with oxygen several times. Fill the bag attached to the inlet with oxygen and check that the other bag is empty. (The application of suction until the bag collapses is the best method of ensuring that it is empty.) Both large taps are in the closed position. When the subject is seated and mouthpiece and nose-clip adjusted, open the tap on the oxygen bag. The subject then inspires oxygen and expires into the atmosphere for seven minutes. At the end of this time a maximum inspiration is taken, the tap to the collecting bag is opened, the subject expires maximally into the bag and the tap is then closed.

Introduce a sample of the expired air into a gas-analyser and estimate the nitrogen content of the sample. This value is normally less than 2.5 per cent. In emphysema values as high as 10 per cent. are obtained because the nitrogen in the less well ventilated alveoli has not been washed out during the 7-minute oxygen-breathing period.

MECHANICS OF BREATHING

In order that the lung volume should increase during inspiration, force has to be applied to the chest by the respiratory muscles. The volume increase depends not only on the force supplied by the respiratory muscles but also on the mechanical properties of the lungs and thorax. The resistance offered by the lungs and thorax to the muscles is due to:

1. Elastic resistance.
2. Frictional resistance to deformation of tissues.
3. Resistance in airways to flow of air.
4. Inertial resistance (small and disregarded).

3.15 Lung Compliance

The elastic resistance of the lungs can be measured if we know the pressure applied to the lungs to make them increase to a known volume. Both the pressure change and the volume change are measured under 'static' conditions i.e. with no air movement occurring. If pressure and volume are measured during breathing, then the pressure change at a given volume will be greater than under static conditions because some of the pressure is needed to overcome non-elastic resistance. However, during breathing in normal persons there are two points when there is no air movement, namely at the end of inspiration and at the end of expiration. So we can make use of these two points in measuring lung elasticity. This relationship of volume to pressure is usually expressed as a volume change in litres per cm of water pressure change. This is called 'compliance' and is a measure of the stiffness of the lungs.

Method. Volume changes are measured by a spirometer fitted with a potentiometer circuit giving an electrical output of volume changes. This is fed into the 'Y' axis of an oscilloscope. As intra-pleural pressure cannot be recorded routinely, we make use of the pressure difference between the oesophagus and the mouth to give us the pressure being applied to the lungs. This is measured by means of a gas-filled latex balloon in the oesophagus and a capacitance manometer$_{26}$, the electrical output of which is led to the 'X' axis of the oscilloscope. The subject applies a nose-clip and then breathes quietly to and from the spirometer. The simultaneously recorded volume and pressure give the pressure-volume loop as seen in Figure 3.5. If points A and B are joined, the slope of AB gives us the compliance. If the angle α is acute then the lungs are stiff; if the angle is large then the lungs are compliant. The work done on the lungs to overcome elastic resistance during inspiration is given by the area ABC. The area AIB gives the work done during inspiration to overcome the non-elastic resistance offered by the lungs. The area AIBC thus gives us the total work done on the lungs during inspiration. The energy stored in the stretched lungs is in part used to perform the work against non-elastic resistance during expiration, given by the area ABE. Widening of the pressure-volume loop is seen in any subject who has an increased airway resistance, for example, a person who suffers from the bronchiolar constriction of asthma and the extent to which the loop has been broadened can give some indication of the severity of the disease.

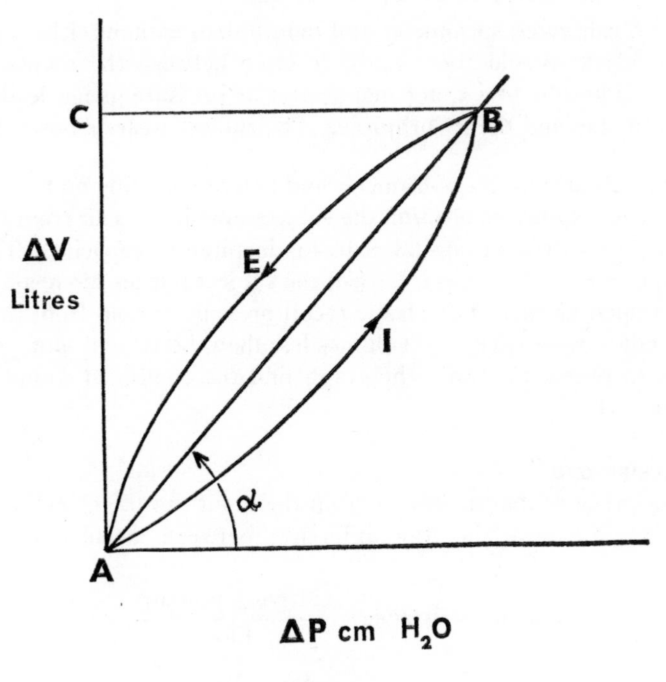

FIG. 3.5

3.16 Combined Compliance of Chest and Lungs

Apparatus: A calibrated spirometer and mouthpiece without either valves or soda-lime absorber is used. A wide bore tap is inserted between the mouthpiece and the spirometer and a side tube to a water manometer or pressure gauge leads off from the airway between the tap and the mouthpiece. The subject wears a nose-clip.

Procedure: Draw air into the spirometer and note the reading on the scale. Starting from his normal end-expiratory position the subject now inhales air from the spirometer till his lungs are quite full, i.e. he takes in his inspiratory capacity. This volume is read from the spirometer. The tap is closed, the subject relaxes his respiratory muscles completely with open glottis. The elastic recoil pressure is read from the manometer. Repeat the procedure with a series of volumes less than the IC and plot a graph relating lung inflation with pressure. From the graph find the combined compliance in litres per cm water pressure.

3.17 Airway Resistance

The driving pressure that moves air from the mouth to the alveoli and back again is the trans-airway pressure, i.e. the difference between alveolar and atmospheric pressure.

$$\text{Airway resistance} = \frac{\text{Driving pressure}}{\text{Flow}}$$

$$= \frac{P_B - P_A}{\dot{V}} \ \text{cmH}_2\text{OL}^{-1}\text{sec}$$

The body plethysmograph is used for the measurement of the pressure difference. When this difference becomes positive, the lungs expand and thus the box pressure rises. During expiration the trans-airway pressure becomes negative, the volume of the chest diminishes and the box pressure falls again. This pressure is displayed on the Y-axis of an oscilloscope and the air flow, as measured by a pneumotachograph, is displayed on the X-axis. The resistance is measured from the slope of the line that appears during normal breathing.

3.18 Gas Sampling

A sample of atmospheric air is easily obtained and a sample of expired air can be obtained by connecting a collecting bag to the expiratory side of the mouthpiece and one-way valve system. The collection of alveolar air is a little more difficult. As expired air is a mixture of dead space air and alveolar air, if we could discard the first part of an expiration, a sample from the second part would give us alveolar air. This holds good only if we are dealing with a normal person with even air distribution in the lungs. The separation of the sample can be arranged if the subject exhales into a 1.3 m long tube. At the end of expiration the air in the tube at the mouthpiece end will be alveolar, and so a sample of alveolar air can be obtained at this point.

Fix a gas sampling tube in a clamp about 1 m above the table on a stand in the middle of a large tray. A mercury reservoir is connected to the lower end of the sampling tube by a length of pressure tubing. Turn the taps so that on raising the reservoir the sampling tube is filled with mercury and the air is completely expelled. Close the upper tap of the sampling tube and lower the mercury reservoir to table level. When the mercury has all run out of the sampling tube close the lower tap and disconnect the reservoir.

Sterilize the mouthpiece of the 1.3 m long tube and attach the sampling tube to the side tube near the mouthpiece. The subject breathes normally for a short time, then at the end of a normal inspiration puts his mouth to the mouthpiece, expires deeply and closes the mouthpiece with his tongue. The upper tap of the sampling tube is opened to admit the air sample and then closed. This is the Haldane-Priestley method.

Alveolar air is saturated with water vapour at body temperature (37° C); the water vapour pressure is about 47 mm Hg. The tension of CO_2 and O_2 in the alveolar air is calculated thus:

$$P_{A_{CO_2}} = \frac{\%CO_2}{100} \times (\text{Barometric pressure} - 47)$$

$$P_{A_{O_2}} = \frac{\%O_2}{100} \times (\text{Barometric pressure} - 47)$$

3.19 End-tidal Sampling

A few respiratory gas analysers work sufficiently rapidly for it to be useful to sample the air in the mouth continuously. An example is the infra-red CO_2 meter. This shows zero CO_2 content during inspiration, but during expiration, once the dead space air has been washed out, alveolar air is being sampled and analysed. Thus the breath-by-breath P_{CO_2} can be measured.

It is possible to collect a sample of alveolar air without the disturbance to the normal pattern of breathing that the Haldane-Priestley method (para. 3.18) entails. In modifications of the Rahn-Otis sampler, the changes in mouth pressure are used to operate a mechanical sampling device so that small volumes of end-tidal air are collected with each breath. Examine the device and analyse a sample collected by it.

3.20 Gas Analysis

For many years the most widely used method for the measurement of respiratory gases has been the Haldane procedure. Many modifications have been introduced but they all use the same basic principle. A volume of gas is measured in a calibrated burette; it is then exposed to potassium hydroxide which absorbs the carbon dioxide and the reduced volume is then measured, so giving the percentage of carbon dioxide originally present. The gas sample can next be exposed to pyrogallic acid which will absorb the oxygen present and the sample volume, now further reduced, can again be measured. The second reduction in volume will give the percentage of oxygen present in the original sample, and the volume remaining gives the percentage of nitrogen.

Using the methods described in paragraphs 3.22 and 3.24 carry out an analysis of:

(a) Atmospheric air.

(b) Expired air.

(c) Alveolar air.

3.21 Respiratory Quotient

Repeat experiment 3.9 and calculate your subject's respiratory quotient as well as his minute volume. The respiratory quotient is the ratio of CO_2 produced to O_2 consumed. It indicates the type of food being metabolized and by applying the nomograms 4.1 and 4.2 the subject's metabolic rate can be calculated.

The composition of atmospheric or inspired air is remarkably constant and values of,

$CO_2 = 0{\cdot}03$ per cent. $O_2 = 20{\cdot}93$ per cent. and $N_2 = 79{\cdot}04$ per cent.

can be assumed. Besides measuring the total volume, V, in the Douglas bag analyse a sample of expired air for percentage content of CO_2, O_2 and N_2. Let these be $F_{E CO_2}$ $F_{E O_2}$ and $F_{E N_2}$.

$$\text{Then, } CO_2 \text{ produced} = \frac{[F_{E CO_2} - 0.03]V}{100} \quad \ldots (1)$$

$$\text{Volume of } N_2 \text{ expired} = \frac{F_{E N_2} \times V}{100}$$

= Volume of N_2 inspired, since this gas is not absorbed or excreted.

But, each 79·04 ml. of N_2 in atmospheric air is accompanied by 20·93 ml. of O_2.

$$\therefore \text{ Volume of } O_2 \text{ inspired} = \frac{F_{E N_2} \times V}{100} \times \frac{20.93}{79.04}$$

$$\text{and volume of } O_2 \text{ expired} = \frac{F_{E O_2} \times V}{100}$$

$$\therefore O_2 \text{ consumption} = \left[\frac{F_{E N_2} \times V}{100} \times \frac{20.93}{79.04}\right] - \frac{F_{E O_2} \times V}{100}$$

$$= \frac{V}{100}\left[\frac{20{\cdot}93\, F_{E N_2}}{79{\cdot}04} - F_{E O_2}\right] \quad \ldots (2)$$

From (1) and (2),

$$\text{R.Q.} = \frac{CO_2 \text{ produced}}{O_2 \text{ used}}$$

$$= \frac{V}{100}\left[F_{ECO_2} - 0{\cdot}03\right] \Big/ \frac{V}{100}\left[\frac{20{\cdot}93\ F_{EN_2}}{79{\cdot}04} - F_{EO_2}\right]$$

$$= \frac{F_{ECO_2} - 0{\cdot}03}{\dfrac{20{\cdot}93\ F_{EN_2}}{79{\cdot}04} - F_{EO_2}} \qquad \ldots(3)$$

To calculate metabolic rate the following steps must be taken:

1. From nomogram 4.1 find the calorific value of 1 litre of oxygen at the R.Q. obtained in equation (3).
2. Express the oxygen consumption in litres per minute at S.T.P.
3. From these values the heat output per minute is derived by multiplication.
4. With the help of nomogram 4.2 express this figure in kilocalories per square metre per hour.

In fact a great deal of the above calculation can be avoided. Thus, Figure 4.1 also includes a nomogram which by-passes the calculation of R.Q. and relates the kilocalorie value of a litre of expired air at S.T.P. to the percentage of oxygen in the expired air. Then, the heat output is obtained by multiplying the number of litres of expired air per minute at S.T.P. by the appropriate kilocalorie value per litre. The nomogram includes a protein correction, assuming that 10–15 per cent. of the dietary calorie intake arises from protein.

3.22 The Lloyd Modification of the Haldane Gas Analysis Apparatus

The use of this apparatus[2] will be demonstrated to you. It will yield accurate analyses of oxygen and carbon dioxide only after you have acquired skill in its use. Make analyses of atmospheric air until a result sufficiently close to 20·93 per cent. for O_2 and 0·05 per cent. for CO_2 is obtained. Then proceed to analyses of gases of unknown composition (Lloyd, B. B. (1958). *J. Physiol.* **143**, 5-6*P*).

3.23 Estimation by the Modified Haldane Apparatus (Campbell's)

This apparatus[21] is now in widespread clinical use. An instruction sheet on the method of using it will be found with the instrument. Get a demonstrator to show you the method before you attempt it yourself.

3.24 Infra Red CO_2 Analyser

An even more rapid method of estimating carbon dioxide is also available, which has the added advantage of giving a continuous reading[18]. It works on the principle that infra-red radiation is absorbed by carbon dioxide in proportion to its concentration. A demonstrator will show it in use.

3.25 Paramagnetic Oxygen Meter

A device for measuring oxygen which is quicker and simpler than the Haldane apparatus is the paramagnetic oxygen analyser[18,23]. The principle on which this operates is the measurement of the magnetic susceptibility of oxygen. Among respiratory gases oxygen is unique in being strongly paramagnetic (attracted in a magnetic field) while the others are slightly diamagnetic (repelled in a magnetic field). The analyser consists of a test body (a dumb-bell-shaped body of two small hollow glass spheres suspended by a quartz fibre between two permanent magnets). The gas to be analysed surrounds the test body. Depending on the difference between the magnetic susceptibilities of the glass spheres and the gas which the spheres displace, the test body tends to swing in or out of the magnetic field. In the simpler, but less accurate model, a beam of light is reflected by a mirror attached to the test body on to a scale calibrated in O_2 percentages and tensions. In the more accurate model, the position of the light spot is restored to zero by the current in another circuit. This current is controlled by a multiturn potentiometer, whose dial is calibrated in oxygen percentage. The air passing into the analysis cell must be dry and filtered, note the coloured silica gel in the inlet tube.

1. Insert the nozzle on the inlet pipe into the gas to be analysed.
2. Squeeze the sampling bulb six times, allowing it to expand fully each time.
3. Press the lamp button and either read scale (in simpler instrument) or restore spot to mid point with potentiometer and read dial.
4. Repeat 2 and 3 until constant value is obtained.

3.26 Dead Space Volume

Bohr developed an equation for the dead space volume from the fact that the total volume of expired gas is equal to the volume of the alveolar portion of gas and the volume of the dead space portion.

Using CO_2 as the gas we get the equation:

$$V_D = \frac{[F_{A_{CO_2}} - F_{E_{CO_2}}]\, V_E}{F_{A_{CO_2}}}$$

Following a normal inspiration, breathe into a small collecting bag, measure the volume of expirate and its CO_2 concentration. Using the value already obtained from analysis of the subject's alveolar gas, calculate his dead space volume with the above formula.

In adults it has been found that the weight in pounds is approximately equal to the dead space volume in millilitres.

3.27 Indirect Method of Estimating Arterial P_{CO_2}

The arterial CO_2 tension (Pa_{CO_2}) serves as an excellent index of the adequacy of pulmonary ventilation, indeed the Pa_{CO_2} is so informative that it would be measured in a wide range of clinical conditions were it not for the technical difficulty of its measurement.

A re-breathing method has been devised, however, using the alveolar P_{CO_2} to estimate the arterial CO_2 tension. The basis of this technique is that if we bring into equilibrium the carbon dioxide in a bag, in the lungs and in the pulmonary capillaries, then an analysis of the CO_2 content of the bag will give us the CO_2 tension of the pulmonary capillary blood and in effect measure Pv_{CO_2}.

Method

1. Fill a 2-litre bag with approximately 1·5 to 2 litres of oxygen.
2. Insert the mouthpiece and apply a nose-clip. When the subject is breathing comfortably to and from the atmosphere, turn the sleeve valve so that he breathes to and from the bag. During the first few breaths, partially empty the bag if tidal volume is less than half the bag volume. Re-breathe for one and a half minutes, then close the sleeve valve with the bag full of gas.
3. Lay aside the bag for two minutes.
4. Re-breathe into the bag again for 20 seconds, then close the valve again with the bag full.
5. Analyse contents for CO_2 using the modified Haldane apparatus (Campbell's[21]). Multiply the barometric pressure minus the water vapour pressure in the lungs (47 mm Hg) by the CO_2 percentage. The answer is the tension of CO_2 in the bag. The arterial CO_2 tension is obtained by subtracting 6 mm Hg from the bag tension.

3.28 Breath Sounds

By means of a microphone and loudspeaker the sounds caused by passage of air into and out of the chest will be demonstrated. Listen with a stethoscope over the trachea for bronchial breathing and in the axilla for vesicular breathing and try to appreciate the differences between the two as described by the demonstrator.

3.29 Respiratory Patterns

As respiratory movements reflect respiratory adjustments, a study of these movements is of interest. A useful method of studying these movements which does not interfere with speech or ingestion, is by means of a stethograph or pneumograph. This consists of a corrugated rubber hose-pipe with a stopper in each end and an air outlet pipe to a recording tambour.

Adjust the stethograph around the subject's chest, connect the side tube to the tambour, adjust the amount of air in the system and the position of the corrugated tube on the chest in order to get the maximum response to respiration. Have the subject sitting with his back to the recording drum and then observe respirations in terms of rate, rhythm and amplitude under different conditions.

1. *Normal Respirations.* Record normal respirations on a drum (speed of about 2·5 cm to each respiratory cycle) for about three minutes. Then run a time-trace on the drum. Note the characteristics of the trace such as the rate, the relative duration of inspiration and expiration, presence or absence of a pause between inspiration and expiration or between expiration and inspiration.

2. *The Effect of Swallowing.* The subject takes a mouthful of water and holds it in his mouth without swallowing, while breathing through his nose. Make a record of normal respiration and at the command 'swallow', the mouthful is swallowed and a mark is made on the tracing. A more striking effect is obtained if the subject drinks a beaker (250 ml.) of water in a continuous swallowing movement. Record the respiratory movements before drinking, during the swallowing, and then in the subsequent half minute.

3. *The Effect of Talking.* Record normal respiration and then get the subject to read a passage from a book, e.g. a speech from Shakespeare.

4. *Abdominal and Chest Movements.* With two stethographs record abdominal and chest movements simultaneously. Attempt to inhibit abdominal movement during respiration.

3·30 Breath Holding

It is assumed that in the trained subject the maximum voluntary inhibition that can be exerted over ventilatory drive remains constant, and thus measurement of breath-holding time is an indication of ventilatory drive. The breaking point occurs when the ventilatory drive exceeds the maximum voluntary inhibition.

1. Sit quietly for three to four minutes, breathing normally, then hold breath for as long as possible. How long can the breath be held? Find the average of three trials.

2. After one or two minutes of forced breathing (try to increase the depth rather than the rate of respiration) find how long the breath can be held.

3. With the nose-clip in position, re-breathe for one minute from a small bag. Then find how long the breath can be held.

4. After two to three minutes quiet breathing, take a deep breath, and at the end of inspiration hold the breath. Is the resulting urge to breathe expiratory or inspiratory? Repeat the experiment but exhale deeply and then hold your breath.

5. Place one end of a short glass tube in water and the other in your mouth, then hold your breath. When breaking point is reached, sip water.

6. Determine the effect of practice on the time for which you can hold your breath. Plot the time against the trial number. How long does any improvement last?

Try to account for all your results.

3.31 Effect of Lung Inflation on Breath-holding Time

Use a spirometer to set the initial volume at ½ litre intervals above the residual volume, or simply start after a full expiration, normal expiration, a normal inspiration and a full inspiration and measure breath-holding time in each position. Repeat each measurement several times. How reproducible are your results?

Does the volume of the chest diminish significantly during a breath-hold? Use a stethograph or a body plethysmograph to assess this.

3.32 Breath-holding during Hypoxia and Hypercapnia

The relative contributions of hypoxia and hypercapnia to the total ventilatory drive can be demonstrated if the initial alveolar tensions can be varied. The simplest way of altering these is by overbreathing or rebreathing from a bag for a variable time before the breath-hold. Draw a graph with the breath-holding time on the Y-axis with over-breathing time as − X and rebreathing time as + X. Repeat the rebreathing part using a bag filled with oxygen and plot these results on the same graph. What is the effect of thus removing the hypoxia?

A more precise way is to measure the CO_2 and O_2 tensions of the alveolar gas obtained by a quick deep expiration at the end of the breath-hold. (Haldane-Priestley sample, see para. 3.18.) Any method of gas analysis may be used. The initial gas tensions can be adjusted by breathing for 5–10 minutes from Douglas bags filled with prepared mixtures of N_2 and O_2 in the range 10 to 100 per cent. O_2. Plot the P_{CO_2} against the P_{O_2} for each alveolar sample and draw the best curve through your points. How do you interpret this relationship?

3.33 Breath-holding after Hyperpnoea and Exercise

Fit a stethograph and record respirations on a slow drum (about two respirations per cm). The subject must not be allowed to see the drum during the experiment.

1. Take a trace of, say, six normal respirations; while the drum is still running, the subject, at the order of the observer, holds his nostrils and stops breathing; after a time it will be impossible for the subject to hold the breath any longer; continue to record respirations till they return to normal. The subject should try to avoid modifying his respirations voluntarily and should keep his mind off his rate and mode of breathing. This also applies to sections 2 and 3 below.

2. Take a short record of normal respirations; swing the lever off the drum and then take a series of very deep respirations for two or three minutes; return the lever to the drum and record the last two or three deep breaths; then hold the breath as long as possible and record the effect as before.

3. Take a short record of normal respirations as a standard. Swing the lever off the drum and do standing running for two to three minutes. Swing the lever back on to the drum, record two or three breaths, hold the breath as long as possible and record the effect.

Run a time-trace. Measure the periods of apnoea, calculate the respiratory rates under the various circumstances and interpret the results.

3.34 The Effect of Forced Hyperpnoea on Spontaneous Respiration

Open the side tube of the stethograph and instruct the subject to take full inspirations as deeply and rapidly as possible. Inspiration only should be forced; expiration is allowed to occur naturally (i.e. not forced). This is continued for exactly one minute when the subject is told to stop and to pay no further attention to his breathing. Immediately clip the side tube of the stethograph and begin to record the respiratory movements. There is sometimes a phase of apnoea and then, when the respiration does begin again, it is usually irregular and often 'periodic', i.e. it alternately waxes and wanes (Cheyne-Stokes respiration). Record any changes in the colour of the subject during the period of over-breathing and during the recovery period. The subject should then describe his sensations during the overbreathing and during the recovery period and a careful note should be made of these. Pay particular attention to the occurrence of tingling or cramp in the muscles and spontaneous twitching of muscles. Discuss your results.

3.35 Effects of CO_2 Excess

Records of the effects caused by breathing air containing an increasing percentage of CO_2 can be obtained with the aid of a spirometer.

The soda-lime tower is removed from the spirometer so that CO_2 will not be absorbed and approximately 4 litres of oxygen is added. The subject, with a nose-clip on, inserts the mouthpiece and breathes to and from the atmosphere until his respiration becomes steady. The tap is then turned so that the subject breathes to and from the spirometer. Take a record for five to six minutes but stop if the panting becomes intolerable. There will be no oxygen lack but a build-up in concentration of carbon dioxide will occur. The subject should describe his sensations. Discuss these and the record you obtain.

3.36 CO_2 Inhalation from Douglas Bag

Fill a Douglas bag with 5 per cent. CO_2, 95 per cent. O_2 mixture from a cylinder. Connect a valved mouthpiece assembly (Fig. 3.4) so that the subject will inspire the gas mixture and expire it into a second empty Douglas bag. Record the respiratory movements with a stethograph and find the average minute volume during inhalation. Compare with your resting minute volume while breathing room air.

3.37 Effects of O_2 Lack

Care should be taken with this experiment demonstrating the effects of oxygen lack and a member of staff should be present as people vary in their sensitivity to lack of oxygen. When cyanosis is seen the experiment is terminated even if little change is seen on the breathing record.

Empty the spirometer of gas, insert the soda-lime canister for absorption of CO_2 and fill the spirometer with room air. Repeat the experiment as in 3.31. As the subject breathes to and from the spirometer, the oxygen content will fall fairly quickly as oxygen is being consumed, but because of the presence of the soda-lime there will be

no accumulation of carbon dioxide in the system. Again, the subject describes his sensations and you should dicuss these and the record obtained.

3.38 The Effects of Combined CO_2 Excess and O_2 Lack

The combined effects of a diminishing concentration of oxygen and a rising level of carbon dioxide on the pattern of respiration can also be shown with the spirometer. The soda-lime absorber is removed and the spirometer is filled with room air. The experiment is then repeated as before. This time the enhanced effect will probably result in making the subject terminate the experiment earlier.

3.39 The CO_2-response Line

If the subject inspires CO_2-enriched air mixtures with a normal oxygen content, and time allowed for a steady-state to be reached (about 10 to 15 min), measurements of the ventilation and the alveolar P_{CO_2} during the steady-state reveal a striking linear relation between them. This CO_2-response is a most useful index by which other respiratory stimuli (e.g. hypoxia, acidosis, hyperthermia) can be assessed.

Procedure: The subject must breathe through a comfortable, low-resistance respiratory valve and be sitting at ease. He should read a book. He should take no part in the administration of the experiment. One of the experimenters should observe the subject throughout the experiment and be responsible for the comfort of the subject. The ventilation is measured by any of the methods in paragraph 3.9 and the alveolar P_{CO_2} by the end-tidal sampler and infra-red CO_2 meter.

The gas mixtures: 2 per cent., 3 per cent., 4 per cent. and 5 per cent. CO_2 in air or oxygen can be made up in large Douglas bags beforehand. Better, they can be made up on the spot from cylinders and an air-blower using a set of flowmeters appropriately calibrated. A flow of at least 3 times the subject's minute volume should be available to him via a small reservoir bag, and the excess blown off into the room. The inspired gas needs also to be humidified and warmed by passing over a water bath at 37° C.

Plot out your results as ventilation against alveolar P_{CO_2} and draw the best straight line through them, extrapolating it to the abscissa. Calculate the parameters in the following equation:

$$\dot{V}_E = X(P_{A_{CO_2}} - Y)$$

where X is the slope of the line and Y the intercept on the P_{CO_2}-axis. These represent the sensitivity and threshold of the CO_2-stimulus respectively, and are known to vary predictably with the P_{O_2}, pH, temperature, etc. of the blood.

3.40 Artificial Respiration

It is obviously most important that everyone should be able to perform artificial respiration efficiently in an emergency. The situations in which it is required are drowning, electric shock, overdose of narcotics or anaesthetics and the inhalation of poisonous or non-respirable gases. The method most favoured now is that of mouth-to-mouth respiration described below.

Mouth-to-mouth respiration is taught with the aid of a practice mannikin. Get a demonstrator to show you the technique. In addition a film may be shown.

The first step in resuscitation of a subject who is not breathing is to provide a clear air passage. This is done by hyperextending the head and pulling up the chin. This ensures that the base of the tongue is pulled away from the posterior pharyngeal wall. The head is maintained in this position throughout the whole procedure.

Next the nostrils are pinched between the finger and thumb, the rescuer opens his mouth widely and places his lips round the victim's mouth making as tight a seal as possible, and the victim's lower lip is pulled down in order to open his lips.

The rescuer then blows air into the victim's lungs until the chest is seen to expand fully. If the chest does not rise, then one looks for the cause of obstruction. Any foreign matter is removed and if none is present, the head should be extended further, the chin pulled again and another attempt made to blow air into the lungs.

When the chest has expanded fully, the rescuer removes his mouth in order to take another breath, and the victim's head is held in the same position so that the elastic recoil of his chest will empty the lungs.

This intermittent inflation is continued at a rate of approximately 15 times a minute until spontaneous respiration is resumed. The rescuer should attempt to use twice his normal tidal volume for ventilation of the subject.

If the colour of the patient's skin does not improve, a check should be made on his circulation. An imperceptible pulse and dilated pupils mean that the circulation is inadequate and external cardiac massage should be started. Pressure with the heel of the hand on the lower end of the sternum, sufficient to depress it about 3 to 5 cm is repeated once a second. This will produce an adequate circulation as long as it is continued. This procedure too should only be practised on the mannikin and will be demonstrated by a member of staff.

CHAPTER FOUR

METABOLIC RATE AND BODY TEMPERATURE

Metabolic rate is the rate of production of free energy in the body. It is assumed that this is all in the form of heat and it is expressed as kilocalories per hour. Since the direct measurement of heat production from the whole body involves very cumbersome and expensive apparatus, this method has been reserved to check the validity of indirect methods. Indirect methods use measurement of oxygen consumption as an indicator of heat liberation.

It is necessary to specify the conditions in which the measurement is made. The Basal Metabolic Rate (BMR) is taken as the heat liberation when the subject lies quietly after a night's sleep and at least 12 hours after a meal; in practice this means a measurement first thing in the morning before breakfast. These conditions usually cannot be met by the members of a practical course and, as a compromise, the experiment is conducted as though basal conditions applied. This introduces the student to the practical details of the estimation and the actual result obtained is not taken too seriously.

4.1 Metabolic Rate Estimation by Open-circuit Method

Expired air is collected over a timed period in a Douglas bag[12,37], its volume is measured and its composition found by analysis. From these measurements the oxygen consumption per minute and the respiratory quotient are obtained. The respiratory quotient indicates the type of fuel being burnt and gives the calorific value of oxygen.

Procedure

Fit the mouthpiece, which has inlet and outlet valves (Fig. 3.4) to the subject. Apply a nose-clip; test for leaks. Connect the outlet valve to an empty Douglas bag and time the collection period necessary to three-quarters fill the bag. Mix the contents of the bag and expel a little air through the side tube to clear it of atmospheric air. Withdraw an expired-air sample either into an evacuated glass gas-sampling tube (see 3.18) or an oiled all-glass syringe (30 or 50 ml.). These samples must be withdrawn within 10 minutes of the end of the collection period, because many bags have a differential leak for oxygen and carbon dioxide. Pass the gas steadily out of the bag through a gas meter, within the permitted range of flow rate for the particular meter used, which is marked on the meter. Find the average minute ventilation rate by dividing the gas volume in the bag by the collection time in minutes. Convert the volume to standard temperature and pressure by Nomogram 3.2. Analyse the gas sample by an accurate method, such as the Lloyd modification of the Haldane gas analysis apparatus (3.22). Calculate the respiratory quotient (3.21) and the oxygen consumption. The R.Q. should be between 0·8 and 0·9. Find the calorific value of oxygen at the calculated R.Q. from Nomogram 4.1. Measure the subject's height and weight and find his estimated surface area with the aid of Nomogram 4.2. Express your answer as kilocalories per sq. metre body surface per hour.

4.2 Estimation of Metabolic Rate by Closed-circuit Method. The Benedict-Roth Apparatus

This apparatus is used clinically where approximate values of the metabolic rate are sufficient. It records the respiratory movements and oxygen consumption graphically. Expired carbon dioxide is removed from the circuit by soda-lime reabsorption. The apparatus, one form of which is shown in Figure 4.3 consists of a light cylindrical metal bell which fits loosely into a narrow space between two concentric cylinders; the water which fills this space forms an airtight seal but allows free movement. The bell is counterpoised by a weight connected to it by a chain which runs over a pulley. In this way the bell can move freely to accommodate the oxygen within it at atmospheric pressure. An ink-writing pen attached to the counterpoise weight records the volume of gas in the bell on a specially printed chart calibrated with horizontal lines for volume and vertical lines for time.

The bell is first filled with oxygen from a cylinder provided with a reducing valve. The subject is then connected to the apparatus by two large-diameter tubes inserted into a mouthpiece. His nose is clipped. In one type of apparatus valves are mounted in the tubes so that the subject circulates the gas through the bell and CO_2 absorbent canister by the movements of his chest. In another type an electric pump circulates the gas through the absorbent and bell and this arrangement dispenses with valves and reduces the work performed by the subject against the airway resistance. If your apparatus has a pump, switch it on at this point.

The respiratory circuit is cut off entirely from the outside air. The bell moves up and down with each respiration and slowly sinks as the oxygen is used up. The rate of fall measures the rate of oxygen consumption. While the apparatus is being prepared the subject should rest on the couch. The operation of two models of this apparatus, the Palmer and the Kendrick, will be described.

The Palmer Apparatus. Note the position of the CO_2-absorber; it may be either a cylindrical canister fitted with a flutter valve which screws into position within the spirometer, or it may be a metal box fitted with a low-resistance flap valve inserted in the air-line between the mouth-piece and the spirometer. In either case fill the canister

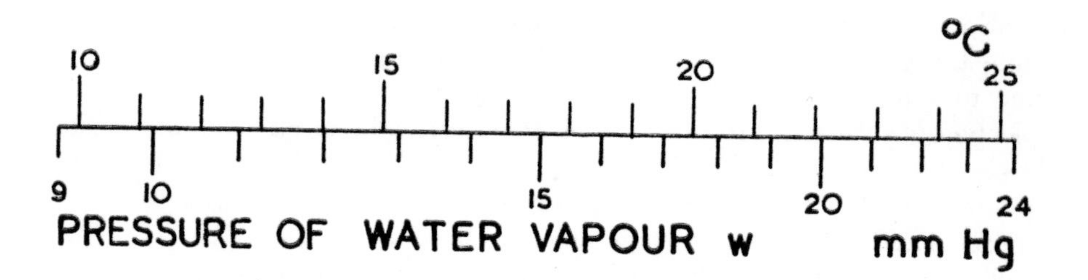

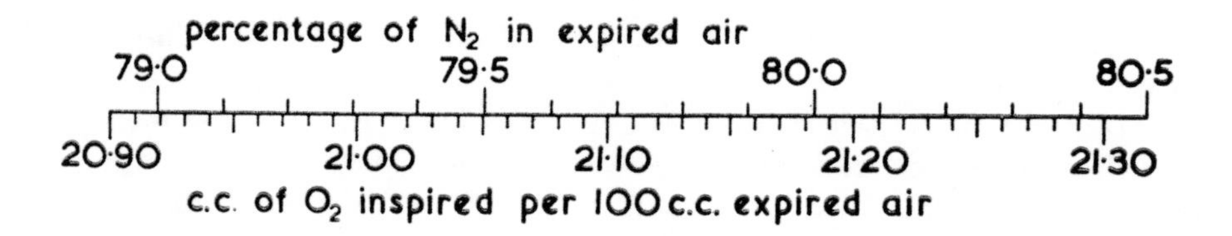

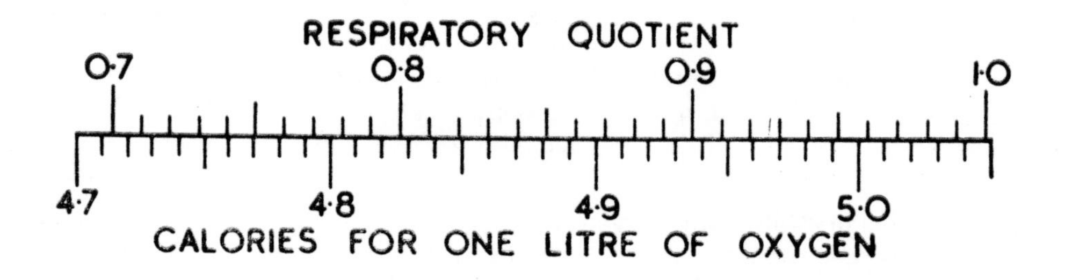

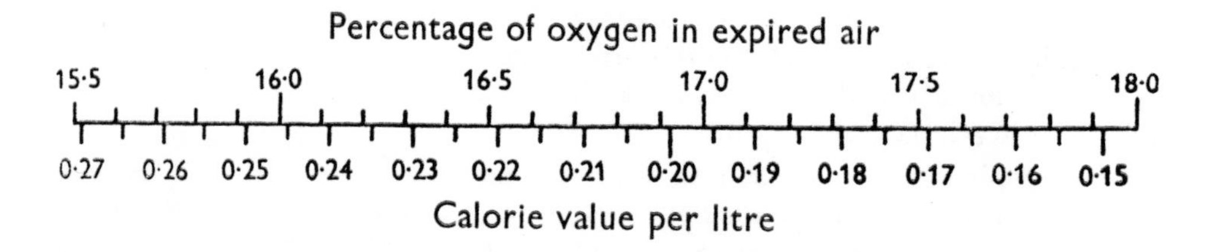

Fig. 4.1

From the percentage of oxygen in expired air, this nomogram gives the Calorie value per litre of expired air. The heat output is given by multiplying the number of litres of expired air by the Calorie value per litre. A protein correction is included on the assumption that 10-15 per cent. of the total Calories arise from protein metabolism. (Weir, J. B. de V. (1949). *J. Physiol.*, **109**, 1). The word Calorie used here is the large calorie or kilocalorie.

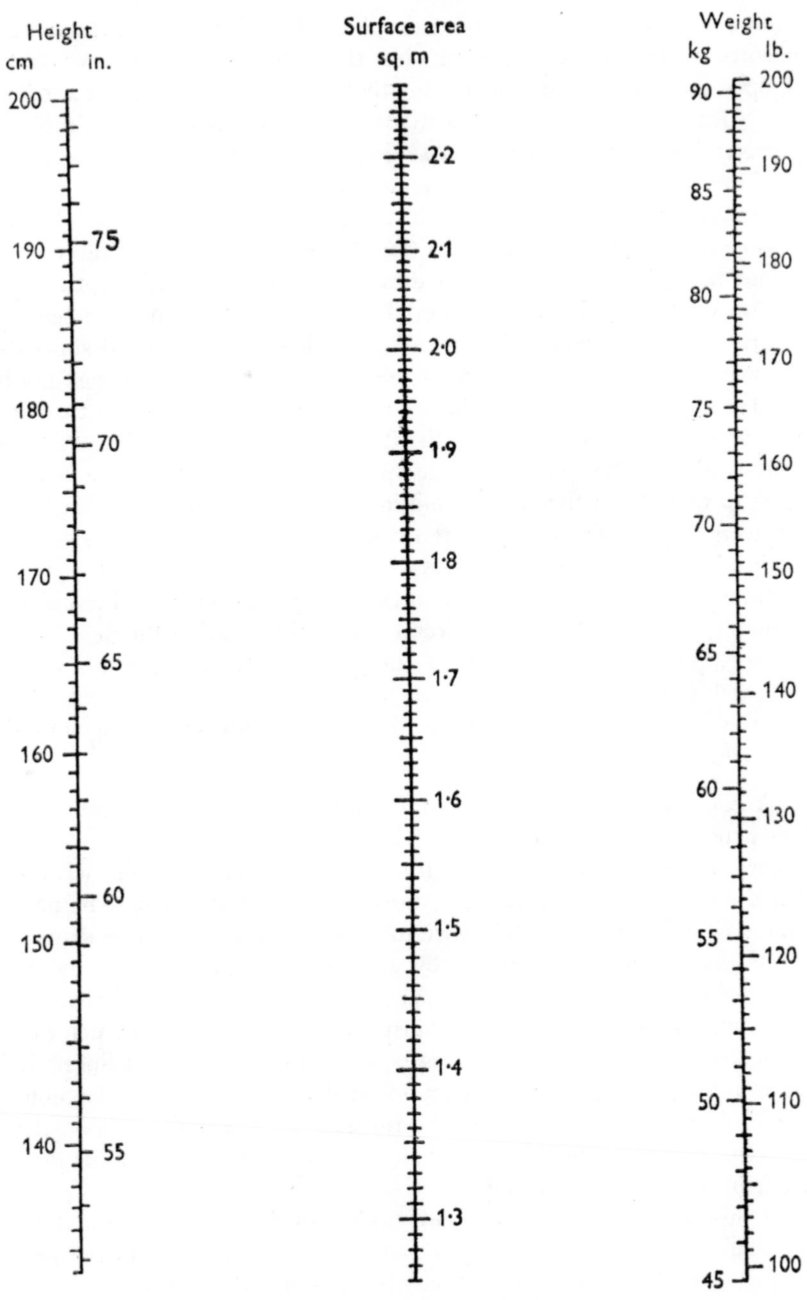

SURFACE AREA NOMOGRAM

FIG. 4.2

(Weir, J. B. de V. (1949). *J. Physiol.*, **109**, 1).

with soda-lime₉. This is coloured green or blue when fresh and changes to orange when its absorbent capacity is exhausted. Fill the outer cylinder with water to create the air seal, open the oxygen inlet tap, and gently push down the bell. Fit a record paper to the drum. Clear the ink-writer with a stilette and fill the reservoir with ink. Fill the bell with oxygen from the cylinder by attaching a rubber tube to the oxygen inlet tap at the base of the spirometer. Close this tap and remove the tubing. Open the mouthpiece assembly valve to atmospheric air. Bring the ink-writer against the drum surface at the zero time line. Test that the ink is flowing freely. The mouthpiece, which is a flanged rubber tube, has been stored in dilute, non-poisonous antiseptic solution, it should now be washed under the tap and fitted to the mouthpiece assembly. The subject, still resting on the couch, is now asked to lie supine, his head supported by a pillow. The mouthpiece assembly is positioned so that the rubber flange fits between the subject's teeth and lips. He grips the rubber blocks projecting from the flange with his teeth and presses his lips inwards on the flange to prevent air leaks. As the mouthpiece assembly valve is open to the air the spirometer does not move and the subject is encouraged to relax for a further five minutes. He should be quite passive and pay no attention to the proceedings. Apply the noseclip to the subject and satisfy yourself that the mouthpiece is correctly positioned. Start the drum motor and close the valve on the mouthpiece assembly at the end of an expiratory movement. The subject's lungs and the spirometer now form a closed-circuit. Sometime during the next five minutes take the subject's pulse at the wrist. When the pen is nearly at the end of the chart, or when the oxygen is almost exhausted, open the valve on the mouthpiece assembly to atmospheric air. Read the air temperature. Release the subject. Swing the ink-writer off the drum and remove the chart.

Kendrick Apparatus. Remove the drum from the drum carrier and wrap a sheet of recording paper around it fixing the overlapping edge with two pieces of gummed paper. Replace it in the drum carrier. Fill the ink reservoir of the ink-writing pointer and see that the pointer writes when it is moved on the paper; it may be necessary to clean out the writing tube with a fine wire to make the ink flow. The drum is driven by an electric clock which is controlled by a switch on the clock. As the peripheral speed of the drum is only about 2·5 cm per minute it is difficult to tell whether it is moving or not unless the pointer is put against the paper. The rate is such that it takes exactly one minute for the pointer to pass between two vertical red lines. It is most important *not* to lift off or replace the drum on the drum carrier whilst the motor (clock) is running. Further the drum must not be turned by hand while it is on the carrier. If it is necessary to alter the position of the drum stop the motor; then lift the drum off and replace it in the required position.

Connect the valves and the large bore tubing to the mouthpiece and spirometer and turn the tap above the mouthpiece so that there is no communication between the spirometer bell and the outside air. Open the oxygen inlet tap and then empty the spirometer by *slowly* and *gently* pressing down the bell. Fill the spirometer with oxygen from the cylinder (see 3.2). While the oxygen is being run in make sure that the

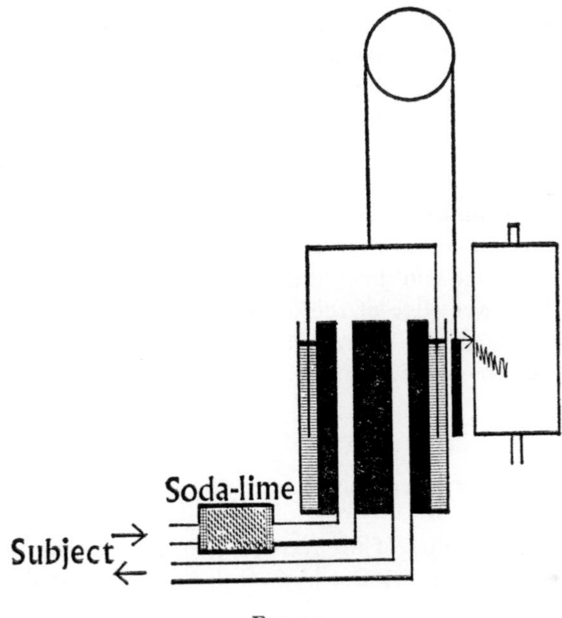

FIG. 4.3

mouthpiece tap is set so that oxygen cannot escape along the inspiratory tube. When the spirometer is full shut off the oxygen cylinder and close the oxygen inlet tap.

The subject should lie down on the couch with his head well propped up. Wipe the mouthpiece with a piece of gauze wrung out in the antiseptic fluid; be careful that no fluid enters the tubes. Place the mouthpiece in the subject's mouth and clip his nose; the subject thus breathes through the mouthpiece to the air. The nose-clip must be properly adjusted since a slight leak through the nose can cause large errors. The subject should be made perfectly comfortable, given a book to read and left undisturbed for at least 10 minutes. When he has settled down turn the mouthpiece tap during inspiration through 90° so that he now inspires oxygen from the spirometer.

The drum is now started and the writing pointer applied. If the pointer is travelling below the level of the lower edge of the drum wait until it has risen above this level before applying it to the drum. Continue the record until the spirometer is nearly empty, then remove the mouthpiece and stop the drum.

Calculations

The same type of record is obtained from the two models and the calculation is identical.

Ignore the first five minutes of the record. During this period breathing is sometimes irregular as the subject accustoms himself to the increased air-flow resistance due to the valves and CO_2-absorber. There is also a phase of equilibration to the high oxygen, low nitrogen composition of the inspired air. Although the spirometer is filled with pure oxygen, nitrogen is contributed from the subject's lungs and body so that pure oxygen is not breathed during the test.

Make your measurements on the later part of the record.

On the chart put: Name of subject, age, height in cm, weight in kg., pulse rate, respiration rate (count from chart), volume of oxygen used in certain time. The rate of oxygen consumption may be found by laying a ruler along the latter part of the record and finding the slope of a line which matches the inclination. If the tidal volume remains constant it is convenient to lay the ruler along the lower turning points corresponding to beginning of inspiration. If the tidal volume is variable draw a line along the mid-points of the tidal excursions. Express the slope of the line in ml. O_2 per minute. A steadily increasing tidal volume may mean that CO_2 is accumulating in the circuit in which case the soda-lime should be replaced. Measure the barometric pressure and the spirometer temperature. Use the Nomogram 3.2 to find the corrected volume of oxygen consumption per minute at S.T.P. Find the volume of oxygen used at S.T.P. in (*a*) 1 hour (*b*) 24 hours. At an R.Q. of 0·82, 1 litre of oxygen at S.T.P. has a kilocalorie value of 4·85. Express (*a*) and (*b*) in kilocalories. Find the body surface area of the subject from Figure 4.2 and calculate kilocalories per sq. metre body surface per hour. Look at Table 4.1 to see the range of B.M.R. values. If your test has been conducted under strict basal conditions, express your answer as a percentage of the normal.

TABLE 4·1

Basal Metabolic Rate

Kilocalories/sq. metre body surface/hour

Age	Male	Female
18	40	36
20	38·6	35·3
25	37·5	35·2
30	36·8	35·1
35	36·5	35·0
40	36·3	34·9
45	36·2	34·5

4.3 The Specific Dynamic Action of Food

The subject should come to the laboratory in the morning without having any breakfast. He rests for half an hour and then his metabolic rate is measured as described in paragraph 4.2. He then eats a good breakfast which has been prepared for him during the metabolic rate estimation. The composition of the breakfast should be decided by the subject. The subject again rests on the couch and his metabolic rate is measured 30 and 60 minutes after the meal. If time permits, further estimations may be made. Calculate the values of metabolic rate as kilocalories per sq. metre body surface per hour and express as percentages of the initial (pre-breakfast) value. Estimate, with the aid of food tables, the composition and kilocalorific value of the meal.

4.4 Metabolic Rate during Work

When measuring oxygen consumption during work it is important that the apparatus does not harass the subject or modify his normal posture and work procedure. Examine the portable respirometer provided (Kofranyi-Michaelis[13]). It is a light gas meter which is accurate over the range of expired air-flow encountered in the working subject and offers little resistance to breathing. It contains a device which samples the expired air throughout the measurement period and collects the sample in a small rubber bladder. A light-weight mouthpiece assembly with inlet and outlet valves is connected to the gas meter by a rubber airway.

Strap the gas meter on the subject's back, fit the mouthpiece assembly to the subject and apply a nose-clip. Allow the subject several minutes to accustom himself to the apparatus before starting the work task. Read the meter and thermometer, start the stopclock, engage the meter recording mechanism and begin the task. Measure the barometric pressure at the end of the task, disengage the meter mechanism, note the time, read the meter, and temperature. Disconnect the sample collection bladder and as soon as possible analyse the expired air by an accurate method, such as the Lloyd modification of the Haldane gas analysis apparatus (see 3.17). Correct the expired air volume to S.T.P. and use the Nomogram 4.1 to obtain the calorific value per litre of expired air. Calculate the subject's metabolic rate in kilocalories per hour.

Body Temperature

Under normal conditions the body temperature of man remains relatively constant within the limits 97° to 99° F, or 36° to 37° C. There is a diurnal variation of about 2° F or 1° C, the maximum occurring in the early evening and the minimum in the early morning. Prolonged muscular exercise may cause a rise of 2° or 3° F but apart from this the temperature in health usually remains within the limits stated.

The temperature may be taken in the mouth, axilla, or rectum. The rectal temperature is the highest (0·5° F above mouth temperature) and is usually regarded as the most reliable. The axillary temperature is not very reliable especially in thin persons. The mouth temperature is reliable, provided that the thermometer is kept there for several minutes. Hot or cold drinks may, however, affect the mouth temperature to a considerable extent.

4·5 Measurement of Body Temperature

The clinical thermometers are left standing in a jar of antiseptic solution (non-poisonous). After use they are washed under the *cold* water tap and replaced in the solution. The glass of the thermometer stem is shaped to act as a convex lens and gives a magnified image of the mercury column. Since this image is visible only from one particular angle the thermometer should be rotated slowly until the mercury comes into view.

1. Hold the thermometer by the end away from the bulb and shake the mercury down. Place the bulb under the tongue and close the mouth. After exactly half a minute take out the thermometer and record the reading. Shake the mercury down again; put the thermometer back in the mouth and let it stay for one minute before removing to take the reading. Continue in this way to take readings after $1\frac{1}{2}$, 2, $2\frac{1}{2}$ minutes and so on until a constant reading is obtained. Plot the results. If it is assumed that the highest temperature reached is a true reading of body temperature how long must the thermometer be kept in the mouth?

2. Take a mouthful of *cold* water, move it well around the mouth and then swallow it. Repeat this several times then take the mouth temperature, leaving the thermometer in for the length of time previously found to be adequate. How does this reading compare with that recorded in (1)?

3. Axillary temperature. Place the thermometer in the axilla and hold it tightly in place for three minutes. Record the reading and account for any difference between it and the mouth temperature.

4.6 The Variation of Body Temperature

You will be issued with a clinical thermometer for a few days. Measure your temperature at least ten times during a 24-hour period; try to obtain several readings during the night. Plot your results on graph paper 1″ per degree Fahrenheit on the vertical axis, 1″ per 3 hours horizontally.

Make observations before, during and after a period (about half an hour) of intense physical exercise. Measure the temperature at the axilla, instead of at the mouth, because the hyperpnoea of exercise will probably produce mouth-breathing and cooling of the inside of the mouth. Plot your results.

CHAPTER FIVE

BLOOD

5.1 The Red Cell Count

Apparatus required: microscope, haemocytometer (counting chamber), red cell pipette, and diluting fluid.

The pipette has a narrow stem, graduated in tenths with figures indicating 0·5 and 1·0, which widens into a bulb containing a red glass bead. The bulb narrows again, and at this point it is marked 101. Beyond this a rubber tube with a glass mouthpiece is attached. Prior to use, the pipette must be clean and dry. Suck up and blow out distilled water several times (fluid flows in and out more quickly if the rubber tube and mouthpiece are transferred to the capillary end of the pipette.) Repeat this procedure with alcohol, then suck up some ether and allow it to flow out under gravity. Finish drying the pipette by attaching the rubber tube to a filter pump; do not blow through it as the moisture content of the breath will prevent drying. The pipette is ready for use when the glass bead rolls about freely with no tendency to adhere to the side of the bulb. The tube and mouthpiece can then be transferred to the correct end of the pipette.

The haemocytometer consists of a thick glass slide with a central platform divided by a short transverse gutter into two parts, each ruled with a counting grid. On each side of the platform, but separated by a trough, there is a support across the width of the slide which holds the special thick coverslip exactly 0·1 mm above the counting area. Both haemocytometer and cover-slip must be dry and free of grease. Clean them with water and then with alcohol. Dry both carefully ensuring that they are free of fibres from the polishing cloth.

Wipe the lobe of your partner's ear with a swab soaked in alcohol and allow it to dry. Hold the ear gently with one hand and with the other make a quick stab into the fleshy edge of the lobe with the disposable sterile lancet$_3$ (Steriseal) provided. Alternatively, draw blood from yourself by stabbing the pulp of a finger or the skin at the root of a nail after prior cleansing with spirit. Wipe away the first drop. Do not rub or squeeze if blood fails to flow freely but wipe the puncture site firmly once more. If blood still fails to flow make a fresh puncture. When a drop of reasonable size has collected, hold the red cell pipette horizontally, apply its tip to the drop and suck up blood to the 0·5 mark or slightly beyond it. Close the mouthpiece by putting the tongue against it. Take the pipette away from the ear and wipe off any blood adhering to its outside. If the blood is beyond the 0·5 mark, tap the tip gently against a finger-nail till the blood is exactly at the mark. Draw a small bubble of air into the capillary.

If there is danger of the blood clotting in the pipette due to delay during these manoeuvres, blow the sample out at once, clean the pipette and begin all over again since pipettes with clotted blood in them are difficult to clean.

Now suck up red cell diluting fluid to the 101 mark. Rotate the pipette vigorously to mix blood and solution thoroughly. This is best accomplished by removing the rubber tube and mouthpiece, holding the pipette horizontally and rolling it with both hands between finger and thumb. Isotonic saline is a satisfactory diluent but special solutions are frequently used. One of these is Hayem's solution which is $HgCl_2$ 0·05 g., Na_2SO_4 2·5 g., NaCl 0·5 g. in 100 ml. of water. This solution tends to cause a deposit in the pipette. A further alternative, is 1 per cent. formalin (v/v) in 3 per cent. tri-sodium citrate. Whichever solution is used it is important to see that no red cells are transferred into it by failure to wipe the pipette clean.

Replace the rubber tube, blow out a quarter of the contents to remove the pure diluting fluid in the stem and lay the pipette down carefully. Prepare the counting chamber by moistening the supports with the tip of your finger and pressing the cover-slip down firmly on them until a series of concentrically-arranged rings (Newton's rings) is seen.

Now touch the tip of the pipette gently on the surface of the counting platform where it projects beyond the cover-glass and a small amount of the solution will be drawn under the coverslip by surface tension. The platform should be completely covered, with no air bubbles or overflow into the central trough. If unsuccessful with your first attempt try again with the other counting area and if this fails, clean the slide and repeat when it is dry.

Place the haemocytometer on the microscope stage and allow two minutes for the cells to settle so that cells and counting-grid are in the same plane. Note that the counting-chamber must be horizontal, i.e. do not tilt the microscope. Scan the counting-area with the × 10 low-power objective together with a small iris diaphragm. If the distribution of cells is not fairly uniform clean the haemocytometer and fill it again.

The counting platform of the Neubauer type which is in general use is engraved with a ruled area measuring 3 mm by 3 mm. This area is divided into nine squares (A) which are again subdivided, the eight outside squares being divided into 16 medium squares (B); the central square is further subdivided into 400 small squares (C). Each of the large squares measures 1 mm × 1 mm and it is therefore 1 sq. mm in area. Each of the medium squares measures 0·25 mm × 0·25 mm, while each of the small squares measures 0·05 mm × 0·05 mm or 0·0025 sq. mm in area. In the *original Neubauer ruling* the 400 smallest squares in the central square millimetre are arranged into 16 groups of 25 small squares each by means of an extra line in the middle of every fifth square. In the *improved Neubauer ruling* a double or treble-ruled boundary line

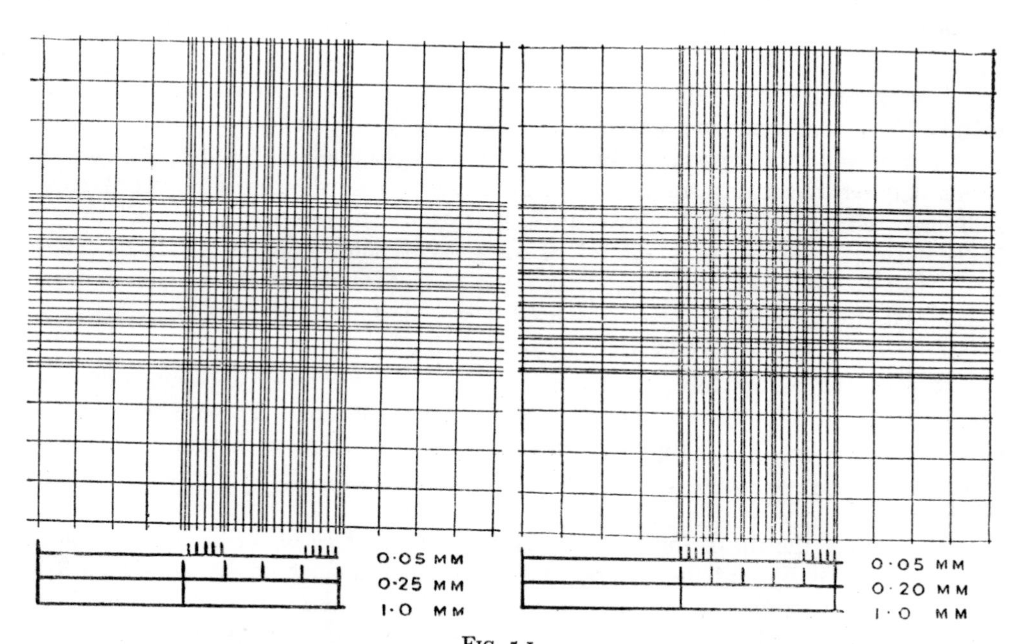

FIG. 5.1

ORIGINAL NEUBAUER IMPROVED NEUBAUER

C in the text refers to the smallest squares (0·05mm × 0·05 mm) seen in the centre of both rulings.

(British Standards 748:1953 has illustrations of all types of haemocytometer counting chambers and dilution pipettes.)

divides the central square millimetre into 25 groups of 16 squares. This means that with the improved ruling the medium squares in the central square only measure 0·2 mm × 0·2 mm but the size of the small squares is the same as in the original ruling, i.e. 0·0025 or 1/400 sq. mm. Since the depth of the counting chamber between the coverglass and the rulings is 1/10 mm, the volume over each of the small squares is 1/4000 c.mm. Bring the ruling into the centre of the field with the low-power (× 10) objective; then use the × 40 (4 mm) 1/6th objective with a small iris diaphragm. It is just possible, if the magnification is correct, to see one group, that is, 16, of the small squares in one field. In order to avoid counting the same cells twice in moving from one square to another and also to decide if a particular cell should be counted, include all cells lying on the upper and left lines of any square, omit the cells on the lower and right-hand lines. Count the cells in five groups of 16 small squares, i.e. 80 small squares, one at each corner and one in the central area. Since the volume of the fluid on each small square is 1/4000 c.mm the total volume is 80/4000. As the blood was diluted 1:200 by means of the pipette the volume of the blood from which the cells came is 80/4000 × 200 = 1/10,000 c.mm. If **n** is the number of red cells found in 80 small squares, then **n** × 10,000 is the number of red cells per c.mm.

Counting chambers of the Thoma type have an engraved area showing exactly the same type of ruling as the central square millimetre of the original Neubauer pattern.

5.2 The Total White Cell Count

The techniques employed are the same as those described in 5.1 and the same counting chamber is used.

The white cell pipette has a smaller bulb than the red cell pipette and contains a white glass bead. The mark above the bulb is designated 11. It requires a larger quantity of blood so do not begin to suck until a large drop is available after pricking the ear or finger. Fill to the 0.5 mark.

The diluent in this case is a 1·5 per cent. solution of acetic acid in water, tinted with methyl violet. The acid destroys the red cell envelopes so that they do not interfere with the count, and the dye colours the nuclei of the white cells making them more prominent. An adequate alternative is 0·1 N-HCl.

After mixing, the pipette is laid aside for five minutes to ensure complete haemolysis. Then the unmixed portion in the capillary tube is discarded.

The counting chamber is filled as described in 5.1, but in this case, using the low-power (× 10) objective, count all the white cells seen over the whole of the ruled area, i.e. nine squares, each 1 sq. mm in area. If the cells are uniformly distributed the figures for total white cells in each sq. mm should not differ from each other by more than eight to ten cells. Since the depth of fluid over the 9 sq. mm area is 0·1 mm, the total number of white cells counted within the 9 sq. mm, **n,** is the number present in 0·9 c.mm. The original dilution of the blood is 1:20; the number of white cells per c.mm is given by **n** × 200 ÷ 9.

One advantage of a slide carrying two counting chambers is that one may be filled with the red cell suspension and the other with the white cell suspension.

5·3 Reticulocyte Count

Two stains are available. Either dissolve 1 g. of brilliant cresyl blue (water-soluble)$_{22}$ in 100 ml. of citrate-saline solution (1 part 3 per cent. sodium citrate with four parts 0·85 per cent. sodium chloride), or, dissolve 0·6 g. new methylene blue$_{22}$ and 1·4 g. potassium oxalate in 100 ml. of distilled water. Both solutions should be filtered before use.

Method. Add 2 drops of the stain to 2 or 3 drops of oxalated blood (depending on the red cell count) in a small (8 mm) glass tube and incubate at 37° C for 20 minutes. Resuspend and spread a film as in section 5.5. Examine successive fields using the 2 mm (× 95) oil immersion objective until the total red cell count includes 100 reticulocytes. The greater the total count the greater the accuracy.

5·4 Eosinophil Count

An increase in the total number of circulating eosinophil polymorphonuclear leucocytes is seen in many conditions, e.g. during asthmatic attacks. Their staining properties can be utilized to assist in counting them.

The stain used consists of 5 ml. of 1 per cent. aqueous eosin Y$_{22}$ and 5 ml. acetone made up to 100 ml. with distilled water. It is only stable for a few days and is prepared freshly for each class.

Make a 1:20 dilution of oxalated blood in this stain using the white-cell pipette. Count in a Fuchs-Rosenthal haemocytometer (the dimensions of this counting-chamber are engraved on it). The eosinophils stain bright red and are easily recognized under low-power (× 10) objective.

Examination of a Blood Film

5·5 Differential White Count

Clean thoroughly three microscopic slides, two to be covered with the blood film and one to be used as a 'spreader'. Ideally the spreading edge should be slightly narrower than the surface on which the film is to be made in order that the edges of the film may be easily examined. A 'spreader' is easily made by cutting off the corners of an ordinary 7·5 cm × 2·5 cm slide with scissors or a diamond.

Clean the ear or the finger with spirit, allow it to dry and then prick it with a disposable sterile lancet to obtain a drop of blood. Touch one end of one of the two slides to the drop of blood—only a small quantity is required. Lay the slide on a flat surface. Place the edge of the spreader on the surface of the slide just in front of the drop of blood and at an angle of 45°. Draw the spreader back until it makes contact with the drop. The blood will now run along the full width of the spreader at the line of junction. When this has happened, push the spreader slowly and smoothly to the other end of the slide. The faster the spreader is moved, the thicker the film; a properly made film should be only one cell thick throughout. Allow the film to dry in air. You can hasten this by waving it about. Repeat with the second slide.

An alternative method of preparing blood films is to use coverslips instead of glass slides. Clean carefully two 2 cm square coverslips: place a *tiny* drop of blood on one coverslip. Hold this coverslip between the forefinger and the thumb of the left hand by two adjacent corners, blood drop upwards. Drop the second coverslip on to the first coverslip so that the corners of the second coverslip project beyond the sides of the first coverslip. Allow the blood to spread out in a very thin film between the coverslips. Immediately thereafter grasp the second coverslip by opposite corners between the finger and thumb of the right hand and *slide* the coverslips apart. Allow the films to dry thoroughly in air. Place both coverslips, film side uppermost, on a single microscopic slide which is resting on a staining rack. See Figure 5.2.

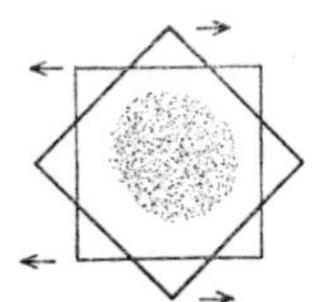

FIG. 5.2

Carefully drop Leishman's stain on to the blood film until the film is covered. Allow the stain to act for one and a half to two minutes. Then add as much distilled water to the stain as will stay on the slide without spilling over. This gives a dilution of 1:1 or 1:2. With a teat pipette suck the diluted stain up and down to mix it thoroughly. If the dilution of stain is correct the fluid will be covered by a thin greenish scum. The diluted stain should be allowed to act for 15 to 30 minutes, then wash it off with distilled water. Make sure that none of the greenish scum settles on the surface of the blood film. Continue washing with distilled water to differentiate until the film has a salmon pink colour. Shake off excess water and allow to dry in air without heat.

For routine examination cover the film with a thin film of immersion oil or liquid paraffin. If a permanent preparation is wanted affix a coverslip with neutral Canada Balsam or DPX mounting medium.

Leishman's stain is a mixture of methylene blue and eosin in methyl alcohol. Preparation of the stain can be complicated and lengthy, so it is usually made up from commercially prepared tablets which contain the ingredients in the correct proportions. In a properly stained film the red cells are pink, the cytoplasm of white cells faintly blue, the small neutrophil granules are purple, the large eosinophil granules brick-red and the basophil granules dark blue.

First examine the film with the × 10 (16 mm) objective: the white cells will be seen to be scattered irregularly throughout the film. Choose an area where there are plenty of white cells and examine these in detail with a 2 mm oil immersion objective and a mechanical stage.

Place a drop of immersion oil on the blood film (with blood films made on slides there is no need to use a coverglass unless a permanent preparation is wanted). Then swing the 2 mm objective into position. Carefully lower the objective until the front lens just touches the drop of oil. Now raise the objective again a distance of 1 mm; capillary attraction will keep the oil in contact with the lens. Look into the microscope and rack down slowly and cautiously with the coarse, not the fine, adjustment until the outline of the cells just appears. Use the fine adjustment for final focusing. If this routine is followed there is no risk of damaging the objective by driving the front lens against the glass slide.

Use the mechanical stage to traverse the full width of the film. Move the slide along horizontally by a distance equivalent to 2 mm and again traverse the full width of the film, this time in the reverse direction. This method ensures that cells are not counted more than once.

Identify each white cell in a series of successive microscopic fields and enter it in the prepared table (Fig. 5.3). Mark only five cells in each vertical column; this maintains a running total of cells examined. The distinction between large and small lymphocytes is rather arbitrary and many regard it as sufficient to group all lymphocytes together. Count 100 cells. Total the table horizontally to determine the percentage of each cell-type in the sample. Compare the values obtained with those listed as 'normal' in your textbook.

A stained film shows the morphological characters of normal red cells and white cells and also the presence of any abnormal types, for example normoblasts, myeloblasts, that may be present. It would also show the presence of certain parasites, for example those of malaria.

The differential count shows any deviation from normal in the relative number of white cells; an abnormal number of cells per unit volume of blood can be shown only by a total white cell count.

5.6 Peroxidase Reaction*

Solutions

1. 40 per cent. formaldehyde.
2. Absolute alcohol.
3. 0·05 per cent. aqueous solution of 2:6 dichlorophenolindophenol (keep stored on ice).
4. 4 vol. hydrogen peroxide.
5. Counterstain, e.g. 0·5 per cent. neutral red.

Technique

1. Fix the freshly prepared air-dried blood film in a mixture of 1 part formaldehyde and 9 parts absolute alcohol for three minutes.

* After Jacoby, F. (1944). *J. Physiol.*, **103**, 25*P*.

Total Count	5	10	15	20	25	30	35	40	45	50	55	60	65	70	75	80	85	90	95	100	%
Neutrophil																					
Basophil																					
Eosinophil																					
Lymphocytes																					
Monocytes																					

FIG. 5.3

2. Wash in water.

3. Treat the film for five minutes with dichlorophenolindophenol solution prepared by adding four drops of hydrogen peroxide to 5 ml. of aqueous 2:6 dichlorophenol-indophenol solution.

The 'peroxidase positive' granules of the myeloid series stain a deep purple violet, most heavily in the eosinophils whose large granules often show a dark purplish peripheral ring and a lighter centre. The monocytes may look as if they have been sprinkled with fine granules.

5·7 Measurement of the Red Cell Diameter

This can be performed in two ways, by diffraction or by direct micrometry. The first is applicable where some accuracy can be sacrificed; otherwise the second method should be used.

If you look at a lamp through an evenly-spread blood film held close to one eye, the light appears to be surrounded by a series of haloes. The size of these is inversely proportional to the size of the red cells. The haloes are produced by diffraction at the edges of the cells. In the method of the Association of Clinical Pathologists, two small lamps are placed 28 in. apart at eye level. When the lamps are viewed with the blood film near one eye (the other being kept closed) two haloes are seen. Each has a yellow centre with a peripheral red rim which is surrounded by a rainbow-like halo with red peripherally. This second red circle is used as the indicator. Starting about 8 ft. from the lamps, move backwards and forwards till these indicating circles (i.e. the second red ones from the centre) just touch. The distance in feet from the slide to the lamps gives the modal mean diameter of the red cells in μ.

Greater accuracy is achieved by direct measurement of at least 100 cells using a microscope, an eyepiece micrometer and a thin well-stained film. This technique is very laborious but less so if an image-shearing eyepiece$_8$ is used.

Estimation of Haemoglobin

There are many methods of estimating the concentration of haemoglobin in blood; they vary widely in ease of performance, complexity of apparatus and accuracy. In general the less complex the apparatus, the less accurate is the method. Most methods involve a comparison of an unknown solution and a coloured standard. Unfortunately each method uses its own standard. Consequently results should be expressed as grams of haemoglobin per 100 ml. of blood, rather than as a percentage. If however the latter form is adopted the standard from which the percentage haemoglobin was derived must be stated.

It is not possible to speak of the normal value of haemoglobin because normality has a very wide range, but the mean value in healthy adult males is 15·5 g. Hb/100 ml. with a range of 13.5 to 18; the mean in healthy adult females is 13·7 g. Hb with a range of 11·15 to 16·4 (15·5 g. Hb per 100 ml. corresponds to 106 per cent. on the Haldane scale). There is also a considerable range in the number of red cells in health, but the figure $5{\cdot}0 \times 10^6$ per c.mm is a convenient average for the whole population. More precisely, adult males average $5{\cdot}5 \times 10^6$ and females $4{\cdot}8 \times 10^6$ per c.mm.

Mass Screening Methods

5.8 Tallquist Scale

A special piece of absorbent paper is touched against a drop of freshly drawn blood. When the glistening appearance has passed off, the tint is compared with a series of colour standards, making a guess at intermediate values if an exact match is not found.

5.9 Copper Sulphate Solutions

The specific gravity (S.G.) of the blood is a function of all its constituents including haemoglobin content. As a first approximation, however, the other variables, for example plasma proteins, can be disregarded; then the haemoglobin content can be estimated from the S.G. In practice drops of fresh blood are allowed to fall from a standard height (1 cm) into a series of solutions of copper sulphate of known S.G. Drop size is unimportant. Each drop is immediately encased in copper proteinate and maintains its coherence for 15 to 20 seconds. The drop expends its initial momentum in sinking 2 to 3 cm; its subsequent behaviour depends on the relative specific gravities of the blood and the solution. It may rise, remain stationary or sink; the S.G. of that solution in which the drop remains stationary is the S.G. of the blood. The haemoglobin is then read from a table.

Use the copper sulphate solutions and the tables relating haemoglobin and specific gravity provided to measure the haemoglobin of blood samples.

Simple Screening of Individuals or Small Numbers

5.10 Haldane's Haemoglobinometer

The apparatus consists of a pipette to measure 20 c.mm of blood, a graduated diluting tube (the volume at the 100 mark is 100 × 20 c.mm = 2 ml.), a colour standard.

In this method the haemoglobin is converted to carboxyhaemoglobin (HbCO) by using the carbon monoxide normally present in coal-gas.

Fill the dilution tube up to mark 10 with 0·04 per cent. ammonia solution. Clean the pipette and obtain a drop of blood using the methods of 5.1. Suck up to the 20 c.mm mark, wipe off any blood on the outside, adjust the volume exactly to the mark by tapping on the nail. Blow the blood gently into the water and suck up and down two or three times to mix thoroughly. Now remove the rubber tube from the pipette and slip the pipette into a rubber tube connected to the coal-gas supply; pass the pipette down the tube until it is just above the surface of the diluted blood. Allow some gas to pass and withdraw the pipette slowly to fill the upper part of the tube with gas; close the end with the thumb and invert several times; finally draw the thumb across the edge so that none of the diluted blood is lost. Repeat the gas saturation twice, using the same thumb and avoiding loss of fluid. All haemoglobin should now be converted into HbCO. Add distilled water drop by drop from a pipette. After each addition mix the contents by (1) inverting the tube, or (2) sucking the fluid up and down a pipette, or (3) stirring with a fine glass rod. The last method results in the smallest loss of fluid, and it avoids frothing. Continue to add water till the tint in the diluting tube is just darker than that of the standard. Compare the tubes against bright diffuse daylight and also while holding them against a sheet of white paper. Read the level of the fluid in the dilution tube. Continue dilution until the tint is just appreciably paler than the standard. Take the average of this reading and the previous one as the correct reading.

The standard in the Haldane apparatus consists of a 1 per cent. solution of blood which had an oxygen capacity of 18·5 ml. per 100 ml. of blood. After dilution it was saturated with CO. This colour standard is stable. On this scale 100 per cent. indicates that the blood contains 14·6 g. Hb per 100 ml.

5.11 Sahli Method

The technique resembles that in the Haldane method except that the haemoglobin is converted to acid haematin instead of HbCO.

Fill the dilution tube to the mark 10 or 20 with 0·1 N-HCl. Add with the same precautions as before 20 c.mm of blood. Mix and allow to stand for the time indicated on the instrument. Add water drop by drop and again take the average of the readings when the tint is just darker and just lighter than the standard. The standard may be a tube containing fluid or a solid glass rod. The equivalent of 100 per cent. on the Sahli scale in g. Hb per 100 ml. of blood varies with different commercial forms of the instrument. The equivalent for any particular instrument is engraved on the stand which carries the tubes.

Accurate Determinations

5.12 Estimation of Haemoglobin by the M.R.C. Grey-wedge Photomoter

Figure 5.4 shows the optical system of the photometer. The diffusing screen at the end may be illuminated by daylight, or by the lighting attachment. The light passing through the apparatus is divided into two symmetrical pathways; in one a glass or perspex cell 1 cm deep is placed. This contains the coloured solution. The other beam of light passes through a segment of an annular neutral grey wedge and then through a compensating cell filled with water. An appropriate monochromatic light filter is fitted over the eye-piece (in the case of oxyhaemoglobin this is Ilford No. 625, Filter No. 2). The observer looking through the eye lens sees two colour fields of different intensity. The grey wedge is rotated until both halves of the field appear equally bright. It is not possible to match an empty instrument; readings should be made between 15 and 200 degrees on the scale. A neutral grey glass standard with its optical density and haemoglobin equivalent (Haldane scale) etched upon it is supplied with the instrument. For example a filter of 0.475 optical density should give a reading of 100, or 14·6 g. haemoglobin per 100 ml. of blood, so that the accuracy of the instrument may be tested

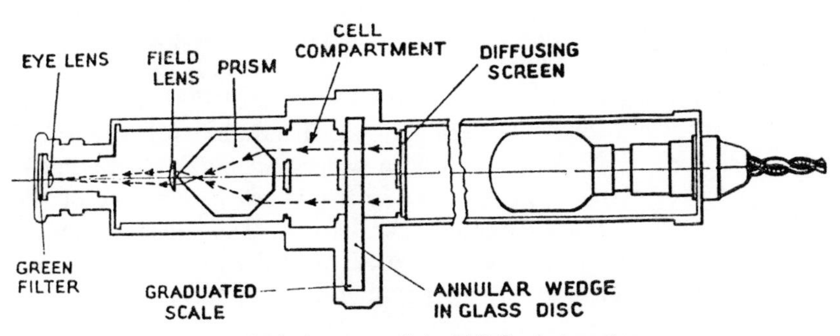

Optical system of the M.R.C. photometer.

FIG. 5.4

King, E. J., Wootton, I. D. P., Donaldson, R., Sisson, R. B. & Macfarlane, R. G. (1948). *Lancet*, **2**, 972.

periodically. The windows of the cell-slot should be wiped clean with a soft cloth each day before use and particularly if solutions are spilt into the cell-slots. The glass cells must be kept clean: if necessary they can be washed in soap and water.

Place about 2 ml. diluting fluid (0·0067N-NH_4OH) in the test tube provided. Using a clean and dry 0·02 ml. pipette, obtain blood from the finger or ear. Any surplus blood is wiped off the pipette and the contents are rinsed into the test tube. Add diluting fluid up to the 4 ml. mark and mix well. This gives a 1:200 dilution of oxyhaemoglobin which remains stable for a few hours. Fill a glass cell and place it in the right-hand cell slot of the instrument. Fill another cell with water and put it in the left-hand slot. Match the two halves of the field by rotating the knurled periphery of the annular wedge. When a match is reached the dial can be read through the small magnifying glass attached to the body of the instrument. The dial reading is the percentage of haemoglobin on the Haldane scale (100 per cent. = 14·6 g. haemoglobin per 100 ml. of blood).

When the estimation is completed the cells should be washed in water and dried thoroughly. The pipette should be washed in water and then in acetone and shaken until dry. A soft towel wrapped round the finger can be used to wipe the glass windows in the inside of the cell slots.

5.13 Photometric Determination

Photoelectric Colorimeters—General Theory

When light passes through a solution of an absorbing substance in a non-absorbing solvent the 'optical density' or 'extinction', E, is directly proportional to the depth of the solution and the concentration, C, of the absorbing substance (Beer's law). E is the logarithm of the ratio of the incident light to the transmitted light. If we compare two solutions equal in depth, $E_1/E_2 = C_1/C_2$. The extinction is determined by measuring the current flowing in a circuit in which a photoelectric cell is connected to a galvanometer, the solution being interposed between a light source and the photoelectric cell. If the galvanometer is calibrated so that its reading at maximum deflexion is zero on a logarithmic scale, the scale readings represent either the extinction E, or E × 100, and solutions of the same pigment give readings which are directly proportional to their concentrations. This method requires monochromatic light obtained by the use of colour filters, the most suitable filter generally being that which gives the maximum absorption, e.g. a yellow or an orange solution requires a blue or violet filter.

When an instrument is used regularly for one particular estimation it is usual to prepare a calibration curve. The absorptions at a series of concentrations of the test substance are measured and a graph constructed; thereafter an unknown solution may be determined at once without the use of standards provided that the same filters, glassware, and reagents are employed.

Instructions for Use of the EEL Portable Colorimeter[16]

1. To protect the sensitive galvanometer movement, the switch on the instrument must always be OFF when the instrument is not in use or when it has to be moved.

2. Close the shutters by rotating the milled wheel in an anti-clockwise direction. Connect to the mains and switch on mains switch and instrument switch; the galvanometer needle should remain at or near infinity, a slight deflection taking place when daylight is admitted as the cover is lifted.

3. Insert the appropriate colour filter in the slot provided. Place the control tube ('blank') in the circular aperture aligning the mark on the tube with the mark on the colorimeter. (The tubes have been matched in this position to eliminate differences in glass thickness). Note that the tubes must be absolutely clean and dry on the outside; any moisture or deposit affects the light absorption and corrosive or fuming substances damage the photoelectric cell.

4. Close the cover. Open the shutter until the meter reads zero. Wait for two minutes and readjust the zero reading which may have drifted slightly.

5. Remove the control tube; insert the tube containing the coloured solution and note the meter reading.

6. The solutions must be completely free from turbidity. Turbidity increases the apparent density and introduces errors.

7. Keep the filters dry and replace them in the box immediately after use.

8. See that the instrument is switched off when your measurements are completed.

Calibration of the Instrument as a Haemoglobinometer

Since the response of the instrument may not be quite linear, it is wise to prepare a calibration curve by using a series of concentrations of a standard solution. The standard solution in this case is an artificial one (Gibson-Harrison)[9] which, undiluted, is equivalent to 16·0 g. of haemoglobin/100 ml. (by iron analysis) (B.S.I. Haldane standard is 14·6 g./100 ml. = 100 per cent.).

Into four test-tubes pipette 2·5 ml., 5·0 ml., 7·5 ml., and 10 ml. of Gibson-Harrison standard. Add distilled water from a pipette to each tube to bring the final volume to 10 ml. Place the tubes in a *boiling* water bath for four minutes, remove, and cool under the tap. Determine the absorption of the cooled solutions (Ilford filter No. 625 and distilled water blank) and prepare a calibration curve relating the colorimeter reading to g. haemoglobin/100 ml. Employ this calibration curve to measure the haemoglobin in each of the samples provided.

Pipette 5 ml. of 0·1N sodium hydroxide into a test tube and add exactly 0·05 ml. of blood; rinse the pipette in the normal way. Place the tube in a boiling water bath for four minutes, cool, and measure the absorption in the colorimeter. Read off the haemoglobin concentration on your calibration curve.

A second method for the photometric determination of haemoglobin is that employing cyanmethaemoglobin. A commercially prepared standard is available, the blood sample being treated with a modified Drabkin's solution.

5.14 Packed Cell Volume (P.C.V.)

This is the volume of cellular elements per unit volume of whole blood. It is usually expressed as a percentage or as a number of ml. per 100 ml. of whole blood. The range for normal subjects is represented by 42 ± 3 per cent.

This ratio of cells to plasma is determined by centrifuging oxalated blood in a graduated tube (haematocrit) until no further reduction in the volume of the cells is noted (usually 30 minutes at 3000 rev/minute). The initial volume of blood and the volume of cells are then read off and the P.C.V. calculated.

5.15 Estimation of Mean Red Cell Volume (M.C.V.)

There is no simple method of measuring the volume of a single erythrocyte directly but the mean cell volume can be calculated from the total red cell count and the packed cell volume.

$$\frac{\text{P.C.V. per litre of Blood}}{\text{Total Red Cell Count in Millions per c.mm}}$$
$$= \text{Mean volume of a single red cell in cubic microns i.e. M.C.V.}$$

The normal range is 86 ± 8 c. μ.

5.16 Estimation of Haemoglobin Content of the Red Cells

Measurements of this are indirect (see 5.15) but the weight of haemoglobin in the average red cell can be derived from the haemoglobin content of the blood expressed as g. per litre and the red cell count in millions per c.mm.

$$\frac{\text{Hb (g. per litre)}}{\text{Red Cell Count (Millions per c.mm)}}$$
$$= \text{Mean Cell Haemoglobin in picograms } (\mu\mu\text{g.})$$

The normal range is 31 ± 4 $\mu\mu$g. Hb per mean red cell.

This expresses the cell haemoglobin content in absolute terms. Alternatively, it can be expressed as a percentage of normal. This is the Colour Index (C.I.).

$$\text{C.I.} = \frac{\text{Haemoglobin (percentage of normal)}}{\text{Red Cell Count (percentage of normal)}} \quad \text{and is,}$$

of course, unity in the 'normal' subject. Its major drawback is evident; it is difficult to agree a normal red cell count and even harder, because of the variety of standards in use, to define normal haemoglobin concentration. Most authorities take 5×10^6 cells per c.mm as the 'normal' count and 91 per cent. Sahli or 105 per cent. Haldane as 'normal' haemoglobin. Any C.I. between 0·85 and 1·15 is accepted as normal.

5.17 Mean Cell Haemoglobin Concentration (M.C.H.C.)

This is the percentage of haemoglobin by weight in 100 ml. of packed cells. The normal P.C.V. is 42 ml. per 100 ml. of blood and the normal Hb is 14·5 g. per 100 ml. of blood, i.e. 42 ml. of packed cells contain 14·5 g. of haemoglobin. By simple proportion, 100 ml. of cells would hold 34·5 g. of haemoglobin in the 'normal' subject. This is the M.C.H.C. The normal range is 35 ± 3 g.; frequently the unit of weight is omitted and it is written as a percentage, thus,

$$\text{M.C.H.C.} = \frac{\text{Hb in g. per 100 ml.}}{\text{P.C.V. in ml. per 100 ml.}} \times 100$$

5.18 Fragility of the Red Cells

You are provided with a rack containing nine test-tubes each 7·5 × 1 cm. See that they are clean and dry. Ensure that the dropping pipette is clean and free from grease; rinse it several times with the reagent to be measured out. Hold the pipette vertically with the third and fourth fingers, leaving the thumb and first finger free to compress the teat; if the drops are discharged at the rate of one a second they will then be all of equal size. Go along the tubes first with 1 per cent. NaCl and then with distilled water as indicated in the table. Afterwards shake the test tubes to mix the solutions.

TABLE 5.1

Tube No.	1	2	3	4	5	6	7	8	9
1% NaCl No. of drops . . .	32	28	24	22	20	18	16	14	12
Water No. of drops . . .	8	12	16	18	20	22	24	26	28
Percentage NaCl . . .	0·8	0·7	0·6	0·55	0·5	0·45	0·4	0·35	0·3

Clean the finger near the nail bed with soap and spirit. Allow it to dry and then prick with a disposable lancet₃. Let a drop of blood fall into each tube. Invert each tube to mix up the cells. Set aside for one hour before inspecting the result. If there is no haemolysis the red cells will be found at the bottom of the tube with clear saline solution above. If some haemolysis has occurred the saline will be tinged red with haemoglobin. If haemolysis is complete the fluid will be uniform in colour throughout and there will be no red cells visible at the bottom of the tube. Record where haemolysis starts (usually 0·45 per cent.) and where it is complete (usually 0·35 per cent.).

As at least nine drops of blood are required for this test it is usually less painful to take blood from a vein by means of a syringe and put it into a bottle with anti-coagulant e.g. calcium oxalate. Drops can then be placed in the test tubes by means of a teat pipette.

5.19 Coagulation Time

Blood is withdrawn from a vein using a disposable sterile 5 ml. syringe₃ and a disposable sterile needle₃ of wide bore, e.g. No. 1. Alternatively if disposable syringes are not available a glass syringe must be sterilized in a hot air oven at 160° C for one hour. A clean venepuncture is essential and any technical difficulty in entering the vein invalidates the result. Four small test tubes, 5 × 1 cm, (Wassermann tubes) are cleaned and rinsed out with normal saline and replaced in the rack provided. The syringe is rinsed out with sterile saline solution and with all aseptic precautions blood is withdrawn from a vein and 1 ml. is placed in each of the four tubes. An assistant starts a stop-watch the instant blood begins to flow into the syringe. Two of the tubes are tilted to 45° at half-minute intervals; when coagulation has occurred the tubes can be inverted without spilling their contents. The remaining two tubes are not disturbed and serve as a check on the end-point observed in the other tubes. The time interval between the moment of withdrawal of blood into the syringe and the moment when the tube may be inverted without spilling the contents is recorded and taken as the coagulation time; it is between 5 and 15 minutes at room temperature.

To eliminate variability due to room temperature the same procedure can be carried out at 37° C. In this case the tubes must be supported in a rack in a water-bath which is kept at this temperature. The blood after withdrawal is placed in these warm tubes with the minimum of delay. The coagulation time at 37° C is about four minutes.

(See 6.41 for Bleeding Time).

5.20 Prothrombin Time

In this experiment tissue extract (usually referred to as thromboplastin) and calcium are added in excess to plasma and the clotting time recorded. In such a system the limiting components are prothrombin and plasma clotting factors necessary for the conversion of prothrombin to thrombin. The clotting time in such a test system is referred to as the prothrombin time. The exact time obtained varies with the source and method of preparation of the thromboplastin. Two methods of preparation of thromboplastin are described below, that from acetone-dried rabbit brain giving a normal prothrombin time of 11 to $12\frac{1}{2}$ seconds and that from fresh rabbit brain giving a time of 15 to 18 seconds.

The reagents used are:

1. *Thromboplastin*

(a) Prepared from acetone-dried rabbit brain as described by Quick. The rabbit brain is prepared by removing completely all blood vessels, that is, stripping off the pia, and then macerating the brain under acetone in a mortar. By replacing the acetone several times, a non-adhesive granular powder is obtained, which is dried on a suction filter. This material is placed in small ampoules which are evacuated for three minutes by means of an oil vacuum pump and then sealed. Such a preparation retains its full activity apparently indefinitely. It is best, however, stored in a refrigerator. Before a test about 0·3 g. of the dehydrated rabbit brain is thoroughly mixed with 5 ml. of physiological saline and incubated at 50° C for 10 to 15 minutes. The coarse particles are removed either by very slow centrifugation or by spontaneous sedimentation and the milky supernatant liquid is the thromboplastin reagent used.

(b) Prepared from fresh rabbit brain. The brain is removed from a freshly killed rabbit and all the membranes and blood vessels carefully removed. The brain is placed in a mortar and 5 ml. of 0·5 per cent. phenol in saline added. With a pestle covered by gauze a fine emulsion is prepared. This is diluted with a further 75 ml. of 0·5 per cent. phenol saline and the thromboplastin is ready for use. It should be kept in a refrigerator but not allowed to become frozen; under such conditions it remains potent for six to eight weeks.

2. *Calcium Solution*

Aqueous calcium chloride, 0·025M is used (0·28 g. anhydrous calcium chloride dissolved in 100 ml. distilled water).

3. *Plasma*

Sodium citrate, 0·5 ml. of 3·8 per cent. is pipetted into a centrifuge tube with a graduation mark at 5 ml. About 5 ml. of blood are withdrawn by clean venepuncture and 4·5 ml. delivered to the graduated tube and immediately mixed with the citrate by repeated gentle inversion. After thorough mixing the tube is centrifuged and the supernatant plasma separated. One to two ml. each of plasma, thromboplastin and calcium solution are placed in a water bath at 37° C, in separate small test tubes. A tube of 8 mm internal bore is held half-submerged in the water bath; 0·2 ml. of plasma is delivered to the bottom of this tube and 0·2 ml. of thromboplastin added. After approximately two minutes 0·2 ml. of the warmed calcium solution is added and the stop-watch started simultaneously. Use a small glass rod with its end turned to stir the mixture until the first threads of fibrin are pulled out. This is taken as the end point.

When this procedure is used in clinical practice to investigate haemorrhagic disease or to control anticoagulant drug therapy, blood is obtained from a normal person and examined at the same time to act as a control. This is necessary because of the variations in thromboplastic potency. The test must be carried out within three to four hours of collection of the sample by venepuncture.

5.21 Thrombotest[24]

There is a commercial reagent available which is specifically deficient in factors II, VII, IX and X. All other factors are held constant and the coagulation time is therefore exclusively dependent on these factors in the blood sample to be tested.

Capillary blood can be used; this eliminates the need for centrifugation. Exact details of the method and a correlation curve for estimating coagulation activity are supplied with each batch of reagent.

5.22 Erythrocyte Sedimentation Rate (E.S.R.)

1. *Wintrobe's Method.* This employs venous blood anticoagulated with a mixture of potassium and ammonium oxalate. Bottles[15] are supplied containing such a quantity of these salts that the addition of a definite quantity of blood (usually 2·5 ml.) to the bottle produces an insignificant alteration in tonicity and hence the size of the cells. If prepared bottles are not available, pipette 0·2 ml. of a solution containing 2 per cent. potassium oxalate and 3 per cent. ammonium oxalate into a specimen tube (10 ml. size), evaporate to dryness in an incubator or low-temperature oven and stopper firmly. The tube now contains 4 mg. solid potassium oxalate and 6 mg. solid ammonium oxalate to which 5 ml. of blood should be added. Blood is obtained by venepuncture (see 5.19) and the required amount added to the bottle which is then stoppered and well shaken. Using a capillary pipette fill up a Wintrobe haematocrit tube with the oxalated blood to the mark at 10 cm and place it in its special stand. Adjust the levelling screws until the spirit levels show that the tube is set vertical.

One hour later read off the upper level of the red cells. The normal fall for men is 0-9 mm and for women 0-20 mm. The rate depends a little on room temperature, high temperatures hasten sedimentation. In anaemia where the number of red cells is reduced the fall may be greater than in normal blood. Corrections for these factors can be made, see Britton (1963), p. 781-783. Collect results from other members of the class for comparison.

2. *Westergren's Method.* This employs sodium citrate as an anticoagulant and requires a rather larger volume of blood. Prepared bottles[15] are available but can be made up as required. Mark a 1 cm test tube with a circle at 1 ml. and at 5 ml. Fill up to the 1 ml. mark with 3·8 per cent. sodium citrate solution. With aseptic precautions take a little more than 4 ml. of blood from a vein. Push the piston of the syringe down slowly and allow the blood to drip slowly from the needle into the tube until the 5 ml. mark is reached. Suck up the blood mixture into a Westergren tube and allow it to empty into the test tube two or three times; finally suck up to the 0 mark. Place the tube in a special stand with the lower end (marked 200) on a rubber bung in the base of the stand; the upper end is fitted into a clip so that the tube is held vertical. At the end of one hour read the lower end of the clear plasma. If the subject is not anaemic the reading should be less than 5 in males, and less than 10 in females.

BLOOD GROUPS

Two main problems occur in blood transfusion. The first is to find if the blood of a potential donor is compatible with that of the recipient (cross matching). The second problem is to find the blood group of a given blood sample (grouping).

5.23 Cross Matching

This is a direct test of compatibility of donor cells and recipient serum. Obtain 1 ml. of blood from the recipient's vein and allow it to clot. When the clot retracts, the serum can be pipetted off. Alternatively several capillary tubes may be filled with blood from a finger prick; when the blood has clotted each tube yields a drop of serum.

Prick the ear or finger of the potential donor. One drop of blood should be placed in 1 ml. of saline and mixed. Alternatively suck up blood to the 0·5 mark of a white cell pipette and dilute to 11 mark with 0·9 per cent. saline, i.e. dilute 1:20. Mix and blow out the saline in the stem. Saline dilution of the cells prevents false agglutination due to rouleaux formation.

Place one drop of the suspension of the donor's red cells on a microscope slide and over it place one drop of the recipient's serum. If the bloods are incompatible this will be shown by an agglutination or clumping of the red cells, readily seen with the naked eye. This may occur quite soon, but may not develop for several minutes if the agglutinin titre of the serum is low. Therefore it is always advisable to wait at least 10 minutes before pronouncing two bloods compatible.

Blood grouping tests should be performed in a warm room, e.g. near a microscope lamp, to prevent the action of cold agglutinins which are occasionally present and may lead to confusion. The drops of serum must not dry up; therefore carry out tests away from draughts or an open window.

5.24 Determination of the Blood Group

1. Obtain a saline suspension of the subject's blood as directed above. Using a skin pencil divide a microscope slide or a white tile into two areas marked A and B. You are provided with samples of Anti-A and Anti-B serum. Place 1 drop of Anti-A in area A and one drop of Anti-B in area B. Use a separate clean pipette for each sample, label these pipettes and use them for no other purpose. On each drop of serum place a drop of the 5 per cent. suspension of the subject's cells. The result should be read at the end of 10 minutes. The group of the subject can then be obtained from Table 5.2 where + indicates agglutination and − means no agglutination.

TABLE 5.2

Group of Subject	Reaction of subject's cells to anti-B (β) =Group A	anti-A (α) =Group B
AB	+	+
A	−	+
B	+	−
O	−	−

Blood grouping can be done either on slides or in small test tubes (Kahn tubes). In hot dry climates, where evaporation is rapid, tubes are recommended. The following technique may be used.

2. In a small test tube (Kahn tube) place about 2 ml. of 3 per cent. sodium citrate, and add a large drop of blood obtained by finger prick. Mix well without shaking and centrifuge at low speed for two to three minutes. With a fine Pasteur pipette remove the supernatant fluid from above the button of cells at the bottom of the tube. Re-suspend the cells in enough normal saline (0·9 per cent. NaCl) to make a 5 per cent.

suspension. Again mix, centrifuge, remove the supernatant and resuspend. You now have a cell suspension free of serum.

Take two clean, dry test tubes, the smallest you can get. Label one 'anti-A' and the other 'anti-B'. Mark them in such a way that they can still be identified after centrifuging. Using a clean Pasteur pipette place, in each, one drop of cell suspension and one drop of normal saline. Now add one drop of anti-A typing serum to the tube marked 'anti-A', and one drop of anti-B serum to the tube marked 'anti-B'.

Mix the contents of the two test tubes by flicking with the finger, cap or plug and leave for two hours. If it is essential to know the result in a shorter time, an effect equal to two hours standing can be achieved by leaving to stand for 10-15 min. then centrifuging very gently at not more than 1000 r.p.m. for 60 seconds.

After either of these procedures, raise the cells from the bottom of the tube by one gentle flick and examine by transmitted light with a hand lens.

If agglutination has occurred the cells are seen in large clumps or even as a single particle. If no agglutination has taken place, you will see a uniform suspension with the cells often rising from the bottom of the tube in smooth swirls. If in doubt, spread one drop of the tube contents on a clean slide and examine under the low-power of the microscope. Determine the group of the unknown cells provided.

5.25 Rh Group using Complete Anti-D

1. Prepare a 2 per cent. cell suspension of the patient's red cells in isotonic saline (2 drops of blood to 2 ml. of 0·85 per cent. sodium chloride solution).

2. Add one volume of the cell suspension to one volume of a known anti-D serum in precipitin tubes (50 × 7 mm round-bottomed and lipless). Stopper the tubes.

3. Incubate for two hours at 37° C.

4. The presence or absence of agglutination may be ascertained by viewing the button of sedimented cells. A negative result is shown by a perfectly smooth-edged small round button and positive agglutination by a larger deposit of cells with irregular or fluffy edges.

5. Confirm the result microscopically. Take great care in removing the agglutinates and spreading them on the slide or they may be destroyed.

Controls. Simultaneously, the following controls may be set up under the same conditions as the test:

1. The Positive Control. A 2 per cent. cell suspension of known O D-positive cells tested against the anti-D serum used. If available, use cells of type CDe/cde and avoid D-positive cells with the E antigen present, as such cells are the stronger D reactors.

2. The Negative Control. A 2 per cent. cell suspension of known O D-negative cells (preferably of genotype Cde/cde) tested against the anti-D serum used. If this is positive, it indicates that the serum is contaminated or that there is an additional antibody present.

TABLE 5.3

SUMMARY OF VALUES

Hb % ≡ g./100 ml.	Total W.B.C. /c. mm
R.B.C. × 10^6 cu. mm	Differential W.C.
Reticulocyte Count	Neutrophil %
C.I.	Eosinophil %
	Basophil %
P.C.V.	Lymphocyte %
M.C.V. c. μ	Monocyte %
M.C.H. μμ g.	
M.C.H.C. %	
E.S.R. Wintrobe mm 1st hour	Blood group ABO
Westergren mm 1st hour	Rh

BLOOD GAS ANALYSIS

5.26 Determination of Oxygen Content and Capacity of Blood by Simplified Haldane Method

The apparatus is shown in Figure 5.5. It consists of a pear-shaped bulb (B) of about 25 ml. capacity fitted with a side tube (S) of about 1·5 ml. capacity, and carrying a bored ground-glass stopper which is connected by rubber tubing to the tap (T). By means of the tap the bulb may be connected either to room air through the side tube (A), or to the gas burette (G), or to both A and G simultaneously. On the right are the positions of the tap which give these various connections. The burette (G) is calibrated in units of 0·01 ml. up to 1 ml. Its lower end is connected to the levelling reservoir (L). This is used to set the level in the burette to any convenient point or to maintain the gas in the burette at atmospheric pressure. The reservoir contains acidulated water (0·5 per cent. H_2SO_4) to which has been added a few drops of methyl red and also a few drops of a wetting agent (such as 'Teepol'—Shell). The whole is mounted on a stand which carries a large water bath into which the bulb can be inserted. A Terry's spring tool-clip is fixed inside the water bath to hold the bulb upright. A thermometer is fitted in the water bath so that its temperature can be measured.

Method

1. Pipette 2 ml. of ammoniated water (0·5 ml. of 0·880 ammonia in 100 ml. water) into the bulb (B).

2. With a 1 ml. pipette, run 1 ml. of blood under the surface of the water.

3. Put the tap (T) into position ⊤, stopper the bulb and immerse the bulb in the water bath.

4. *Wait for three minutes.*

5. By means of the levelling tube (L) set the level of the liquid in the burette to some conventional value near the lower end of the scale. Turn the tap to position ⊣ and read the level indicated on the burette. Call this reading (a).

6. Shake the bottle vigorously for half a minute so as to lake the blood and allow it to absorb oxygen, then return the bottle to the water bath and *wait for three minutes* before proceeding further.

7. Re-level the liquid menisci in the burette and levelling tube (so as to bring back the gas in the bulb and burette to atmospheric pressure), and then read the new volume on the burette. Call this reading (b).

8. The difference between readings (a) and (b) is the quantity of oxygen absorbed by 1 ml. of blood. This is a measure of the degree of unsaturation of the blood.

9. Turn the tap to position ⊤, take the bulb out of the water bath and remove the stopper from the bulb.

10. Using an angled 1 ml. calibrated pipette, put 0·5 ml. saturated potassium ferricyanide solution into the side tube. Use great care to ensure that none enters the laked blood solution.

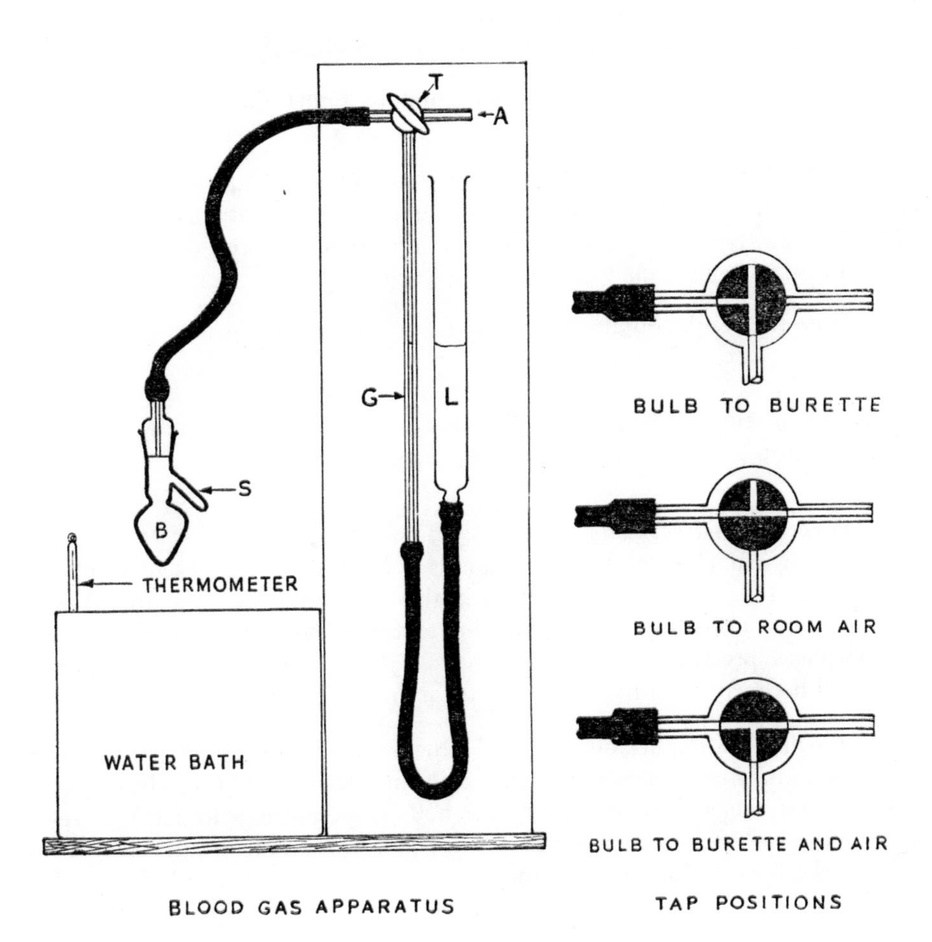
T
A
G
L
S
B
THERMOMETER
WATER BATH
BULB TO BURETTE
BULB TO ROOM AIR
BULB TO BURETTE AND AIR
BLOOD GAS APPARATUS
TAP POSITIONS

FIG. 5.5

11. Replace the stopper, put the bulb back into the water bath and *wait for three minutes.*

12. Set the liquid meniscus to a convenient level near the top of the burette, turn the tap to position ⊣ and read the level. Call this reading (c).

13. Shake the bulb vigorously for half a minute, return the bulb to the water bath, and then *wait for three minutes.*

14. Re-level the menisci and read the new gas volume. Call this reading (d). The difference between (c) and (d) indicates the volume of oxygen driven off from 1 ml. of blood by the potassium ferricyanide. This is a measure of the oxygen content of oxygen-saturated blood (usually called the oxygen capacity).

15. From the results obtained in §8 and §14 calculate the original oxygen content as follows:

Oxygen Capacity − Oxygen Unsaturation = Initial Oxygen Content.

Express all three values in ml. oxygen per 100 ml. blood.

Notes on Apparatus and Method

1. The original Haldane blood apparatus contained a compensatory tube which was connected to a second bulb immersed in the same water jacket as the blood bulb. This compensated for changes in volume due to alterations in water bath temperature, or in atmospheric pressure, during an estimation. In fact, neither change is very likely to occur, and both can be corrected for by taking readings of the factors concerned.

2. A much more important source of error is the change in volume of the gas in the bulb due to the heat generated by friction during shaking. This can be compensated for by allowing the gas to cool to its original temperature before taking readings. This is the reason for the insistence on a three-minute wait at certain times during an estimation. Perform a dummy experiment using only water in the bulb, shaking for half a minute, and then reading the gas volume in the burette immediately and at 15-second intervals after shaking, for a period of three minutes. Plot the results as a graph (volume against time) to show the need for this precaution.

3. Another source of error is in the use of saponin as a haemolysing agent. Many samples absorb considerable quantities of oxygen. For general purposes ammoniated water alone gives adequate results. A borate buffer can be used instead of the ammonia solution. The composition is 4 ml. 0·1 N-NaOH + 6 ml. NaH_2BO_3 solution (NaH_2BO_3 solution contains 12·404 g. boric acid and 100 ml. N-NaOH in 1 litre). The pH of this buffer is 10. When using this borate buffer solution the most satisfactory haemolysing agent is 10 drops of Johnson's Photographic Wetting Agent No. 326, added to the buffer.

5.27 The Natelson Microgasometer[17]

This instrument measures the gas content in very small quantities of blood, adapting to ultra-micro analysis the classical Van Slyke manometric method. Accurate results are obtained on a test specimen of only 0·03 ml. blood. The main use of the apparatus is in estimation of CO_2, O_2 and CO.

A detailed description of the technique involved accompanies the apparatus.

5.28 Carbonic Anhydrase

Red blood corpuscles contain an enzyme, carbonic anhydrase, which catalyses the reaction $CO_2 + H_2O \rightleftharpoons H_2CO_3$ and enables it to proceed fast enough in either direction for efficient gaseous exchange to take place in the lungs and in the tissues. This experiment demonstrates that the release of CO_2 from a solution of bicarbonate by weak acid is accelerated by carbonic anhydrase and that the enzyme is inhibited by cyanide, a powerful general anti-enzyme.

Measure 1 ml. of bicarbonate solution (0·2M, 1·68 g. per 100 ml.) into the flask of the simplified Haldane blood-gas apparatus (see 5.26) and put 0·8 ml. of acid phosphate solution into the side-arm. (Acid phosphate: equal parts of Na_2HPO_4 2·84 g. per 100 ml. and KH_2PO_4, 2·72 g. per 100 ml.). Close the flask tightly, open the tap, and wait for three minutes. Raise the levelling tube until the meniscus in the burette stands exactly at zero and close the tap.

Ideally three students should conduct this test since the reaction is over very quickly. Student A is the shaker. After starting the reaction he holds the flask in the water-bath and shakes it regularly, twice a second, in time with a watch. Uniformity of shaking rate in repeat experiments is important as changes may affect the rate of evolution of carbon dioxide.

Student B reads the burette. Removing the levelling tube from its clips, he lowers it as carbon dioxide is evolved, so that the menisci in the burette and levelling tube are always at the same level; i.e. the gas is always at atmospheric pressure. Student C is time-keeper and recorder.

Student C gives the word 'Go' and starts the stop-watch. Student A tips the phosphate into the bicarbonate, immerses the flask in the water-bath and starts shaking. *Every five seconds* student C calls out the time, student B instantly reads the burette (remembering to level up) and calls out the reading to C who records it. This is continued until four successive readings have been the same.

Empty, clean and dry the flask.

Measure 0·02 ml. of blood with a haemoglobinometer pipette from a finger prick and eject it into 4 ml. of distilled water: this is a 0·5 per cent. solution.

Repeat the experiment, *exactly* as above, adding 0·1 ml. of the 0·5 per cent. blood solution to the bicarbonate in the flask.

Finally repeat it again, adding to the bicarbonate in the flask, 0·1 ml. of the blood solution and 0·5 ml. of the cyanide solution (KCN, 0·008 g. per 100 ml.).

You now have three sets of volume readings. Plot them on a graph, time on the abscissa and volume on the ordinate.

5.29 The Preparation of an Oxygen Dissociation Curve

1. The sample of oxalated blood to be used is exposed to the desired degree of vacuum in a tonometer or sampling bulb for half an hour (Fig. 5.6).

2. A piece of thin rubber tubing is fixed to the outlet of the flask and the rubber bung at the flask's inlet removed.

3. The tap controlling the outlet of the flask is opened and the blood allowed to drip slowly out of the end of the rubber tube.

4. A hypodermic syringe, lubricated by two to three drops of liquid paraffin, and freed of air is used to collect the sample as it flows down the tube. The tube is treated as a 'vein' and the needle of the hypodermic syringe is pushed into it.

5. The blood is then transferred from the syringe to the blood gas analysis apparatus (5.26 or 5.27).

6. Construct an oxygen dissociation curve by plotting the partial pressure of oxygen for different samples of blood against the oxygen content as determined by the blood gas analyser. The partial pressure can be calculated from the reading of the manometer. Allowance should be made for the pressure of water vapour at the temperature used.

5.30 Flame Photometric Estimation of Sodium and Potassium in Serum

Metallic salts ionize when drawn into a non-luminous flame and emit light of a characteristic wavelength.

In the EEL Flame Photometer$_{16}$, coal-gas and air flow into a mixing chamber at controlled rates, the airflow at the point of entry being used to draw the sample solution through a fine atomiser jet into the same chamber. The mixture is ignited and the light produced is passed through a narrow-band optical filter which transmits only the wavelength peculiar to the metallic cation under study. These filters are readily interchangeable. The filtered light energizes a photoelectric cell and the output of the photocell is applied to deflect a galvanometer. By varying the sensitivity, the instrument will measure concentrations over a wide range but the most accurate results ($\pm$ 1 per cent.) are obtained with dilute solutions such that full-scale deflexions are given by 5 parts per million of sodium or 10 parts per million of potassium.

The galvanometer scale is linear and readings are proportional to the current generated in the photocell but the relationship between flame intensity and solution concentration is not linear. Calibration curves are, therefore, constructed using a series of standard dilutions of each cation.

With mixed solutions of ions interference effects occur but at the concentrations present in plasma after dilution these are a negligible source of error.

Procedure: You are supplied with suitable calibration curves of sodium and potassium and standard solutions which correspond to 100 on these curves, namely sodium, 1 mg./100 ml. and potassium 1 mg./100 ml.

Prepare a 1:500 dilution of the plasma sample in distilled water to measure sodium and a 1:50 dilution to measure potassium.

Now set up the photometer.

1. Turn the 'sensitivity' control fully anticlockwise.

2. Select and insert the optical filter appropriate to the cation under test.

3. Switch on, ensuring that the galvanometer is unclamped.

4. Turn on the gas supply, open the gas control and light the flame.

5. Turn on the air supply and adjust the air control to give a reading of 10 Lb./sq. in on the pressure gauge (or as specified with the instrument).

6. Slide a beaker of distilled water into the sample recess and raise it to the stop.

7. Adjust the gas control to produce a flame with one large, central, blue cone then slowly close it down until 10 separate cones form. This standardizes the flame.

8. Replace the distilled water by the appropriate standard solution and adjust the sensitivity to give approximately full-scale deflexion.

9. Return to distilled water and set 'zero'.

10. Present the standard solution again and set sensitivity so that this solution produces *exactly* full-scale deflexion.

11. Check 'zero' with distilled water.

12. Present the unknown solution and note the reading.

After obtaining several readings for one ion, change the optical filter and repeat the steps from (6) to (12) for the other.

Then switch off the instrument as follows:

1. Turn off gas and pause till flame is extinguished.

2. Turn off the air.

3. Switch off.

4. Clamp the galvanometer.

Calculation: Average the readings obtained for each ion and read off the concentration equivalent to this reading from the calibration curve. The concentration in mg./100 ml. is given by this value × the dilution factor.

Results: Express the sodium and potassium concentrations in serum in mg./100 ml. and in m-equiv/litre (A.W. of Na = 23; A.W. of K = 39·1). The latter is more commonly used in medical practice. Note that calibration curves and standards can be prepared so that the photometer readings can be equated directly with m-equiv/l. e.g. if a potassium standard containing 9·8 mg.K/litre is employed to set full-scale deflexion, the instrument will read over the range 0-0·25 m-equiv/l. which is 0-12·5 m-equiv/l. of undiluted plasma. This more than covers the range compatible with life.

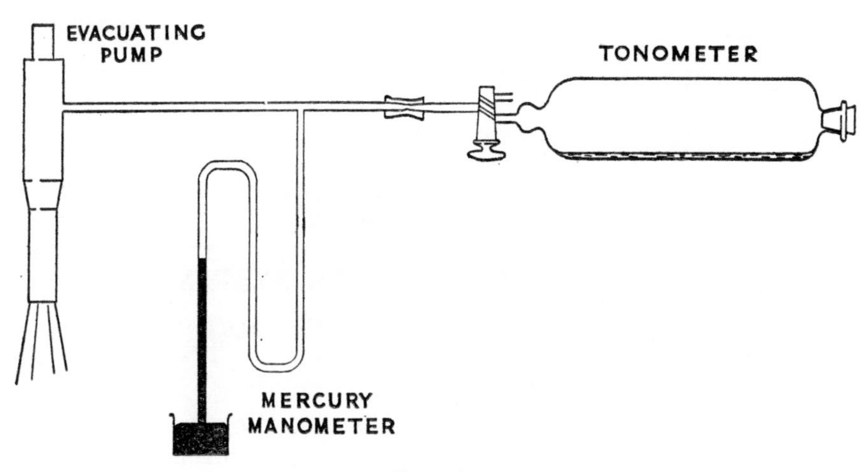

Fig. 5.6

CHAPTER SIX

THE CARDIOVASCULAR SYSTEM

CIRCULATION IN THE FROG

6.1. The Cardiac Cycle of the Frog

Apparatus required: frog bath, pillar on adjustable base, frog heart lever, drum, needle and cotton thread. Study the anatomy of the heart as demonstrated by the large plasticine models. Have all the apparatus ready and the drum smoked before beginning the experiment.

The frog is killed and divided into an upper and a lower portion as described on p16. Pin the upper part to the cork board ventral side up. Pick up the abdominal wall with forceps and cut in the mid-line till the sternum is reached. Pick up the sternum with forceps, cut on either side with scissors, keeping the points well up and pointing slightly laterally. Cut through the shoulder girdle and the base of the flap containing the sternum. Moisten the heart with Ringer's solution. Usually the pericardium will be cut and it can easily be pushed over the heart. If not, pick up the pericardium, not the heart, with the forceps and make a little snip with sharp scissors. Enlarge the incision in the pericardium from the apex to the base of the heart. Transfer the preparation to the organ bath with the exposed heart directly below the clip which is hanging from the recording lever clamped to the pillar. On pressing the upper part of the clip the jaws open; release the pressure and allow the clip to grip the extreme apex of the ventricle. The ventricle must not be gripped between the forceps during this manoeuvre or at any other time. The surface of the heart must be kept moist. Do not fill the organ bath with Ringer's solution; instead use a pipette and teat to flush the surface frequently with fresh solution and leave the bath drain open. Drying of the heart is shown by disappearance of its surface sheen and this will ruin the preparation.

Gently raise the lever on the pillar till the thread is just right, and cut through the frenulum. This is a fold of pericardium on the dorsal surface of the heart which passes from the ventricle to the venous end of the heart and tends to restrict the movement of the recording lever. The thread should be vertical and sufficiently tight to give a good movement of the writing point; the mean position of the writing lever must be

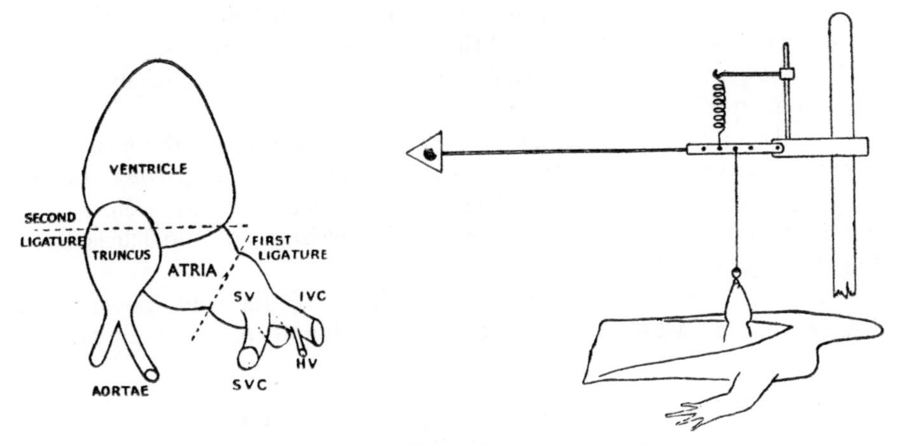

FIG. 6.1

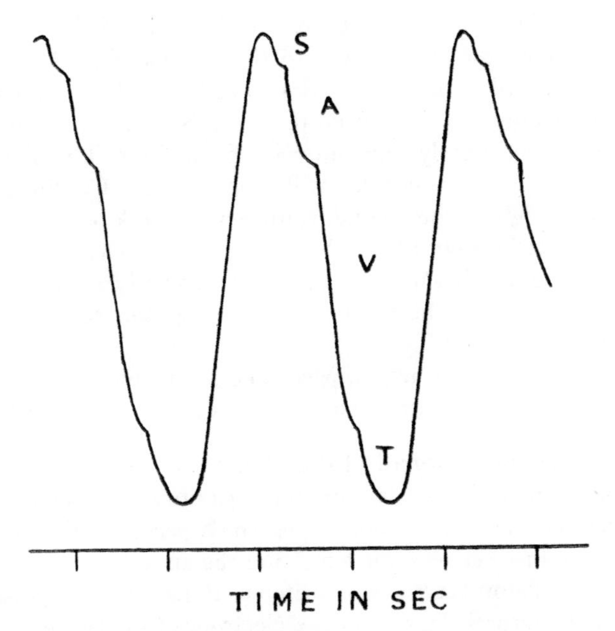

FIG. 6.2

S, A, V and T indicate the waves produced by the contraction of the sinus, atria, ventricle and truncus respectively. The waves are seen most clearly with a large vigorously acting heart, but even a small heart, if carefully handled, will show the A and V waves.

horizontal. The best position of the lever spring on the lever and the amount of tension exerted depends on the size and vigour of the heart. Adjust the points of attachment to the lever of the heart-clip thread and the lever spring so as to obtain the maximum excursion of the pointer on the smoked drum. Use a drum speed of about 2 cm per sec. Set the time marker at one second.

Observe the order of contraction of the various chambers of the heart and label the corresponding waves on the trace. To accomplish this, one partner watches the heart and calls out 'Atria' and 'Ventricle' when the chambers are seen to contract; the other partner observes the trace as it is being written out and labels the waves accordingly. See Figure 6.2.

Note that the trace is the algebraic summation of all the contractions, so that if one chamber is contracting while another is relaxing then the lever may remain stationary.

6.2 Initiation of Contraction. (Stannius Ligatures Experiment)

Set drum speed at 0·5 cm per sec. While the heart is suspended vertically pass a thread by means of a needle or a fine forceps between the aortae (nearer the head) and the veins (caudad) (Fig. 6.1). Tie the first part of a reef-knot *loosely* over the line dividing sinus from atria (the crescent). Tie a similar loose knot around the vertical thread to the recording lever, and slide it down till it lies just below the shoulder of the ventricle. Adjust the position of the lever so that a good record is obtained *without stretching the heart.* Record a few contractions; with the drum still running tighten the first ligature to separate the sinus from the atria. See that the knot lies actually on the crescent, and does not merely separate veins from sinus. The sinus may continue to contract but the atria and ventricles will stop beating. Continue recording and, after 10 or 15 seconds, tighten the second ligature with a jerk to separate the atria from the ventricle. The ventricle should begin beating slowly at once or after a few seconds delay (Fig. 6.3). If the heart is left undisturbed after the first ligature is tied the ventricles will, sooner or later, begin to beat once more; tying the second ligature initiates ventricular activity much sooner.

Discuss the initiation of contraction, pacemaker, heart block, and the differences in fundamental or natural rate of contraction of various parts of the heart.

6.3 The Nerve Supply and Control of the Heart

In the frog the origin of the accelerator and vagal fibres is in the medulla. The vagal and sympathetic nerves join to form one nerve which passes to the heart and enters it at the white crescentic line between the sinus and the atria.

Turn on your stimulator, set Selector switch to PULSE OFF, set Pulse Rate switch to 20 pulses per sec and Strength to 5. Connect electrodes (bipolar silver-wire electrodes separated by 3 mm mounted on insulated handle) to output terminals. The heart should be set up as in experiment 6.1; record normal contractions with drum speed of 1 cm per sec. Either hold the pin electrodes very gently so that the wires lie across the crescent of the heart without disturbing the action of the heart, or use the clamps provided to

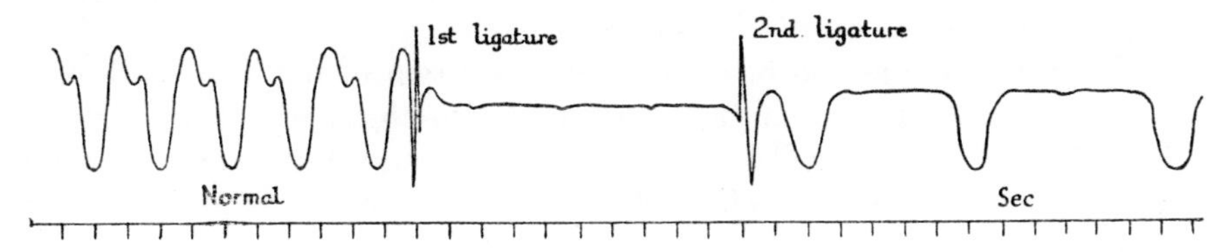

FIG. 6.3

position the electrodes on the crescent. Make a mark on the drum surface a few centimetres forward of the writing point and when the drum has moved round to this point move Selector switch to REPEAT PULSE. After three seconds move back to PULSE OFF and make another mark on the drum. See Figure 6.4.

Usually, during the period of stimulation the action of the heart is disorganized and after the period of stimulation the heart stops due to the effects of stimulation of the vagal fibres. After a period it will begin to beat slowly and weakly and gradually return to normal rate and force. Try the effect of varying the strength and rate of stimulation. Can you show acceleration?

You should now do experiment 6.4 and after the heart has been atropinized repeat the electrical stimulation to see if slowing still occurs.

6.4 Effect of Adrenaline and Acetylcholine on the Heart

Use a drum speed of 0·5 cm per sec. Record a few normal heart beats and then from a teated pipette run a single drop of acetylcholine in Ringer's solution, concentration 2×10^{-4}, over the heart. When an effect has been obtained wash off the acetylcholine with Ringer's solution. The heart beat will return to normal in a short time. Repeat the procedure, but when the heart has been stopped by the acetylcholine, apply atropine sulphate (concentration 5×10^{-4}) in Ringer's solution. The heart will restart: test it with acetylcholine once more. Does it still respond to this substance?

Record the effect of applying a drop of adrenaline, concentration 5×10^{-4} to the heart.

6.5 The Effect of Temperature on the Frog's Heart

Take a record of the cardiac movements in a preparation set up as in section 6.1. You are provided with Ringer's solution at 0° C and at 27° C. While recording, pipette the cold Ringer over the heart and continue to do so until no further slowing is obtained. Now rewarm the heart with room-temperature Ringer. When the rhythm has stabilized apply Ringer's solution at 27° C. (Check the temperature before application, if too hot the heart will be damaged). Use a one-second time trace. Measure amplitudes and rates at the three temperatures. A more elaborate and satisfactory demonstration of the effect of temperature on the cardiac output and rate is described in section 6.15 using the perfused frog heart. An electrically heated wire may be used to explore the heart. With this it may be shown that changes in rate with local heating can be produced readily in the sinus region alone.

6.6 Refractory Period of Cardiac Muscle

Expose the heart and suspend it in the usual way; make records on the drum at a peripheral speed of 2·5 cm per sec. Arrange a signal (i.e. a time-marker) so that it writes about a quarter of an inch below the heart trace but in the same vertical line as the writing point of the heart lever. Wire the signal in circuit as shown in Figure 6.5. Pressure on the push button closes the two independent circuits simultaneously, one

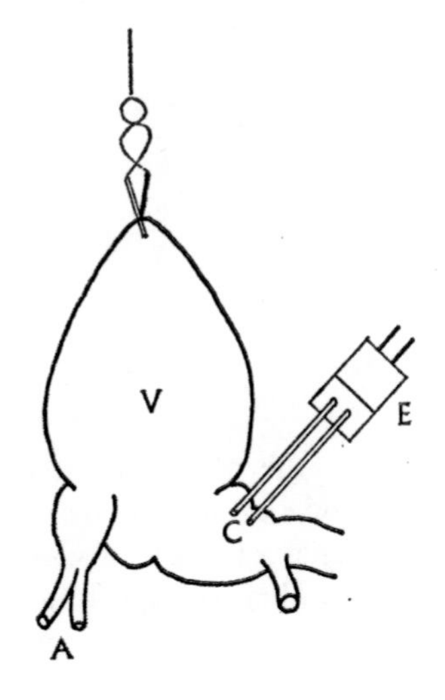

FIG. 6.4
Electrodes E, applied to the crescent C of the frog's heart. V, ventricle; A, aortae.

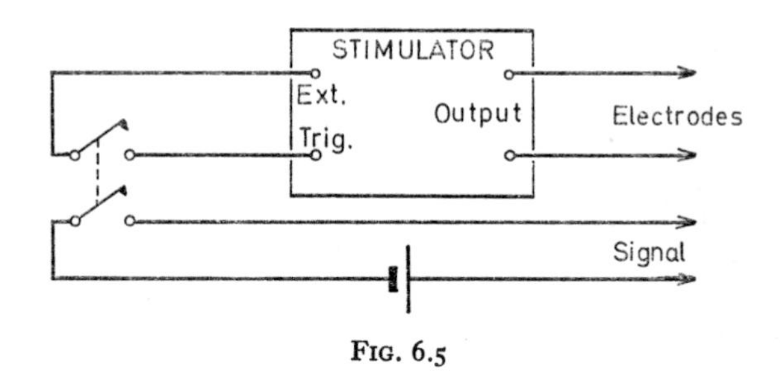

FIG. 6.5

circuit initiates a single pulse from the stimulator, the other marks the moment of stimulation on the drum. Loop one silver wire electrode around the base of the ventricle and attach the other to the metal heartclip on the apex. The electrodes are connected to the output terminals of the stimulator. Adjust stimulus strength to a value sufficient to cause a ventricular contraction during diastole. Now watch the heart closely. Send in a shock during ventricular systole, then a few beats later during early diastole, then in late diastole. Note the pause after each extra systole. Put on 1 second time-trace. Examine your records and decide what is meant by the refractory period of cardiac muscle. Estimate its duration.

6.7 Mechanical Sensitivity of Cardiac Muscle

Suspend a frog's heart in the usual way and tie the first Stannius ligature (para. 6.2) to bring the heart to a standstill. Stimulate the ventricle by scratching it gently with a needle. The scratching will disturb the lever and thus mark the moment of stimulation. Is the ventricular response instantaneous or does it follow after a delay?

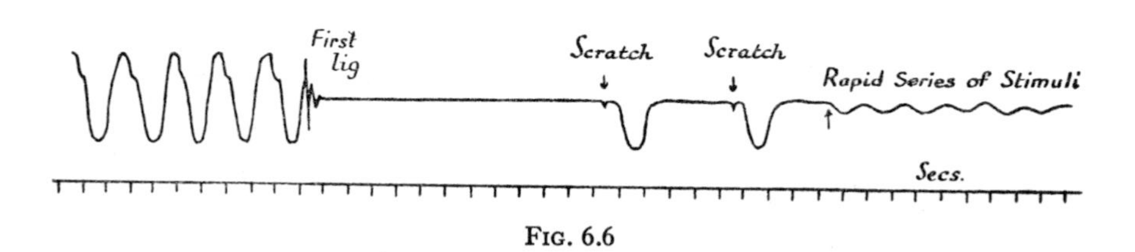

FIG. 6.6

6.8. Latent Period of Cardiac Muscle

Use the same preparation as in paragraph 6.7 but instead of mechanically stimulating the heart apply single electrical stimuli through wick or wire electrodes applied to the ventricle. Use the circuit described in paragraph 6.6. Stimulate the ventricle and measure the latent period.

6.9

Use the same preparation and stimulus arrangement as in 6.8. Switch Selector of the stimulator to REPEAT PULSE and observe what happens when the ventricle is subjected to a rapid series of stimuli. Try various stimulus rates. Compare the responses with those of frog skeletal muscle as in paragraph 2.9.

6.10 The Automatic Action of the Frog's Heart

Expose a frog's heart. Count the heart rate in beats per minute. Cut the heart out of the body, keeping the scissors well away from the bulbus and sinus but avoiding the gall-bladder. Put it in a watch-glass containing Ringer's solution and observe the heart through a low-power (×10) binocular microscope. Note the sequence of contraction of the various parts of the heart, and again count the heart rate. Cut through the sinu-atrial junction, count the rate of the sinus contractions and compare with that of the remainder of the heart. Cut through the atrio-ventricular junction just above the atrio-ventricular groove and report on the behaviour of the various parts. Separate the lower two-thirds of the ventricle. Report on its behaviour and the effect of stimulation, both electrical and mechanical.

6.11 Perfusion of Frog's Blood-Vessels

Use a Mariotte's flask (see 6.13) connected by rubber tubing to a small cannula. Fill the Mariotte's flask with Ringer's solution after clamping the rubber tubing. Insert the stopper carrying the central glass tube into the neck of the Mariotte's flask. Stun and decerebrate a frog but do not pith it (see p. 226). Secure the frog to the frog board which is designed so that the fluid which has perfused the frog's circulation is canalized and may be measured by drop-counting or volumetric collection over timed periods.

Expose the heart by removing the sternum. Pass a thread under one of the aortae and tie loosely the first half of a reef-knot. Adjust the perfusion system so that the supply is free of air bubbles and a slow drip leaves the tip of the cannula when it is held at the level of the frog's heart. Open the aorta proximal to the ligature and tie the cannula into the distal end. Tie a ligature round the other aorta. Cut both atria with scissors to allow the perfusate to escape. Raise the Mariotte's flask to 40 cm above the cannula. Record the perfusion rate.

The effect of adrenaline and acetylcholine can be shown by perfusing the preparation with Ringer's solution containing the drugs in a concentration of 10^{-6}; or small quantities of more concentrated solutions (1 ml. of 10^{-5}) can be injected by means of a syringe through the rubber tubing just above the cannula. After investigating a drug, perfuse with Ringer's solution for some time to get back if possible to the original rate of flow. Find the effect of injecting 1 ml. of 0·5 per cent. sodium nitrite. As the experiment proceeds fluid will escape from the blood vessels into the tissues which will become swollen and tense, i.e. oedematous. The perfusion rate will become less and less and the preparation will become useless for the demonstration of the action of drugs. If the preparation is still in good condition pith the animal and note if this produces any difference in the flow. Has adrenaline any effect now? Describe the action of the various drugs and say whether their action is local or central. What is the cause of the oedema, and why does it slow up the flow?

6.12 Observation of the Circulation in the Frog's Foot

The frog is anaesthetized by an injection of 5 per cent. urethane solution into the dorsal lymph sac, dosage 2 mg. urethane per gm frog weight, with supplements as necessary. Lay it out on a flat piece of cork with one foot over a 1 cm diameter hole near one end. Keep the frog skin moist. Spread the web of the foot over the hole and place the cork board on the stage of a binocular microscope. Examine by transmitted light at magnifications of ×25 and ×50.

Describe the movements of the blood corpuscles in the arteries, veins and capillaries. Are pulsations visible in any of the vessels?

6.13 Perfusion of the Frog's Heart

Introduction

This is an instructive preparation because in addition to the record of the movements of the heart it is possible to record cardiac output. Thus the basic relationships between venous pressure, arterial pressure and cardiac output can be measured. Further experiments show the effect of adrenaline, acetylcholine, temperature and variations in the composition of the perfusion fluid on cardiac output and rhythm. The basic set-up will be described first, and this should be arranged before the dissection is begun. Modifications necessary for particular experiments will be described in the relevant sections.

The Ringer's solution is supplied by a Mariotte flask, which maintains a constant perfusion pressure as the reservoir empties. The flask is mounted on a platform fitted to a vertical rackwork so that the perfusion pressure can be varied at will by changing its height above the frog's heart. The Ringer's fluid is led into the heart by a Greene's cannula. This device is shown in Figure 6.8. It has a vertical tube which acts as a manometer to indicate venous pressure and a side arm with a clipped outlet to permit fluid to be moved quickly through the connecting tubing.

Cardiac movements are transmitted to a spring lever mounted above the heart. Cardiac output is measured by drop-counting or volume collection from the aortic cannula.

Dissection

The largest available frog should be selected. Stun the frog, cut off its head and pith it. Expose the heart as described in section 6.1. Place a magnifier or low-power dissecting microscope over the preparation. Good lighting is essential. Examine the heart, identify the two aortae. Tie off the right aorta and place a loose ligature around the left. Fit the heart clip to the extreme tip of the ventricle and by gentle traction headwards expose the venous vessels of the heart. Cut the frenulum. Pass a loose ligature round the vena cava near the entry of the hepatic veins. Now clamp the Greene's cannula in position so that the tip is just above the sinus venosus. Cut open the vena cava at the point indicated in Figure 6.7, insert the tip of the cannula and tie it in. The pressure in the cannula should be arranged so that Ringer's solution drips out during the insertion of the cannula and washes away the blood which escapes from the cut vena cava. The atria will now fill with Ringer's solution. Adjust the venous pressure to about 0·5 cm water. Cut the left aorta distal to the loose ligature and after the blood has been washed out insert the arterial cannula (1 mm glass tube) and tie it in. The aortic cannula is connected by a slender flexible tube to an outlet where output can be counted or measured. The height of the outlet above the heart will determine the pressure in the aorta. Attach the heart clip to the lever and position the kymograph. Arrange the time marker to write below the heart trace.

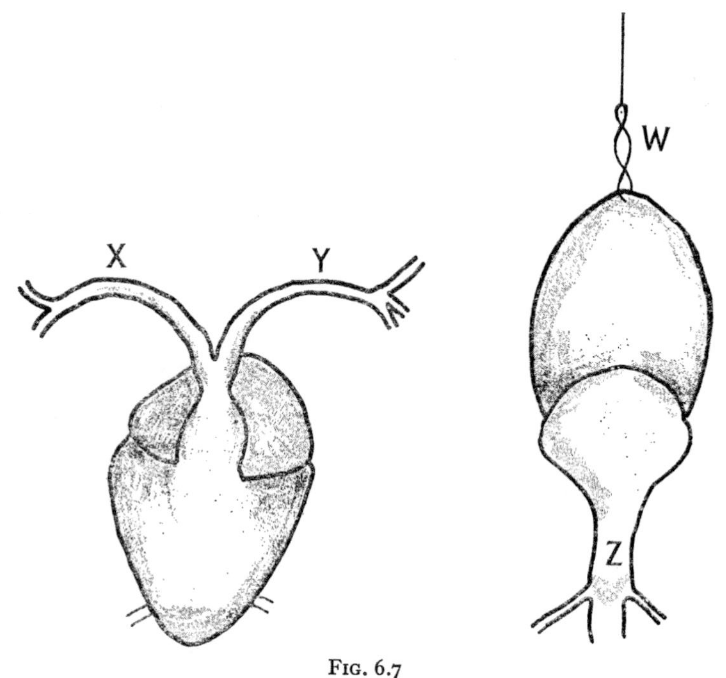

FIG. 6.7
Tie off the right aorta at X, then put a loose ligature at Y. Attach the heart clip, W, and turn the heart forward to expose the vena cava. Open the vena cava at Z and tie in the tip of the Greene's cannula. Section the aorta at Y, and tie in the arterial cannula.

6.14 To Show the Effect of Venous Pressure on Cardiac Output

Record the cardiac output, counting the number of drops in a half minute, over a range of venous pressures. Begin at 0·5 cm pressure and increase in increments of 0·5 cm up to 4 cm or more but use judgment to avoid damage by overstretching the atria. Keep the position of the outflow point of the arterial cannula constant. Bring the pressure down again and record output at a few intermediate points. Plot pressure against output in drops per minute on graph paper. Does the frog's heart give evidence for Starling's Law?

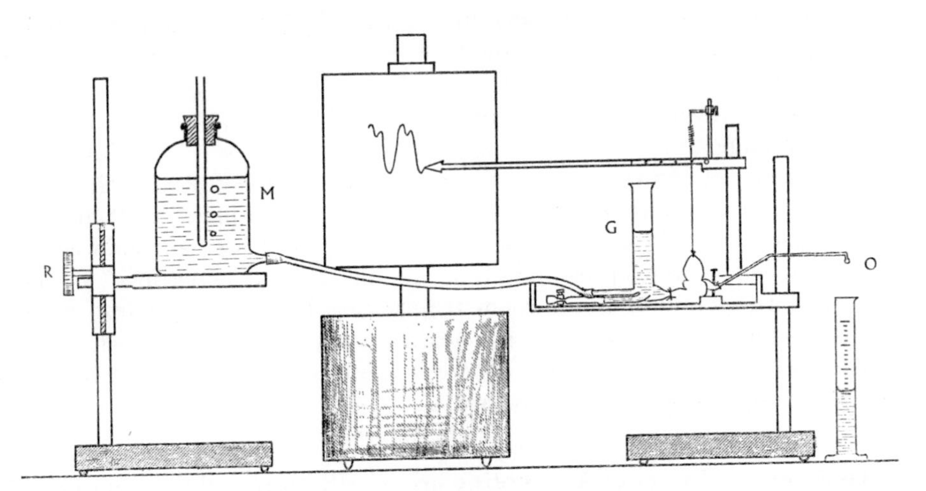

FIG. 6.8

Apparatus for the perfusion of the frog's heart. R, is a rackwork to vary height of Mariotte flask (M). G, Greene's cannula, is tied into the vena cava. O, outlet from arterial cannula: an electric drop-recording device can be placed at this point. The body of the frog is not shown.

6.15 To Show the Effect of Temperature on the Rhythm and Output of the Heart

Introduce a heat-exchanger, in the form of a coil of glass tubing immersed in a beaker, between the Mariotte flask and the Greene's cannula. Place the tip of a thermistor (semiconductor temperature-sensitive resistance) under the sinus venosus and secure in position. The temperature of the thermistor 6,19 is indicated by the current flowing through a microammeter (see Fig. 1.5 for circuit). Set the venous pressure at a value which will give an output of about 40 drops per minute, i.e. 1 to 2 cm pressure. Pour cold water into the heat-exchange beaker, wait until the heart temperature has stabilized and then count output. Use several progressively colder water-baths down to 0° C. Then try the effect of temperatures above room temperature but do not exceed 27° C. Plot cardiac output against temperature, also heart-rate and amplitude against temperature. Consider the consequences to a poikilothermic animal of having a cardiac output which is dependent on environmental temperature.

6.16 To Show the Effect of Adrenaline on the Rhythm and Output of the Heart

(a) Set the venous pressure at 2 cm. Establish a baseline for the outflow at this pressure by counting alternate half minutes for two minutes. Now add a single drop of 10^{-4} adrenaline solution to the vertical pipe of the Greene's cannula and stir with a slender glass rod. Mark the kymograph drum. Record output every half minute until baseline is regained. Plot output, heart rate, and amplitude against time. If this does not produce a marked rise in output, repeat at a higher venous pressure.

(b) A second Mariotte flask containing adrenaline, concentration 5×10^{-7}, in Ringer's solution is set up so that it exerts the same perfusion pressure as the flask containing normal Ringer's solution. A tap permits either solution to be switched to the cannula. Obtain baseline readings and then switch to the adrenaline solution. Record the response and plot output as in (a). Return to normal Ringer's solution.

6.17 To Show the Effect of Changes in the Ionic Composition of the Perfusion Fluid on the Rhythm and Output of the Heart

(a) *The Effect of Calcium Lack.* A second Mariotte flask containing calcium-free Ringer's solution is set up so that it exerts the same perfusion pressure as the flask contanging normal Ringer's solution. A glass tap permits either solution to be switched to the cannula. Obtain baseline readings for alternate half minutes for two minutes, switch to the calcium-free solution and mark the drum. Measure output during the ensuing minutes and when the heart has almost stopped, switch back to normal solution and record the recovery phase. Plot the output, heart rate, and amplitude against time.

(b) *The Effect of Sodium Lack.* Fill the second Mariotte flask with Ringer's solution in which 7 per cent. sucrose has been substituted for the sodium chloride. Repeat procedure as in (a).

(c) *The Effect of Potassium Lack.* Fill the second Mariotte flask with potassium-free Ringer's solution. Repeat the procedure as in (a).

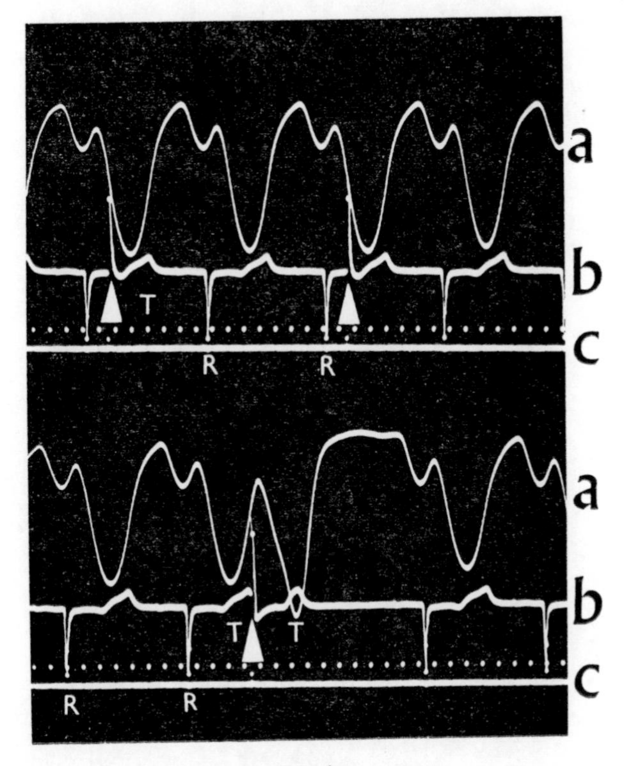

FIG. 6.9

Records of movement and electrical activity in a spontaneously beating frog's heart. a, movement record, systole downward, recorded by a photoelectric transducer attached to the heart lever: b, cardiac electrogram from wick electrodes applied to the ventricle, the R and T waves are shown: c, time marker 0·2 sec, with arrowhead at point of stimulation of the ventricle through another pair of electrodes. The record was photographed from the face of a cathode-ray tube. In the upper set of records a stimulus was delivered to the ventricle at each of the two points marked by an arrow on the time-trace line. It occurred 0·3 sec after the preceding QRS complex and before the associated T wave, and whilst the ventricle was contracting. It had no effect, showing that the ventricular muscle was refractory.

In the lower set of records a stimulus delivered 0·8 sec after the QRS occurred, on the descending part of the T wave during early diastole. This produced an extra systole together with an extra T wave. There was then a pause before the next spontaneous beat.

6.18 The Electrical Activity of the Frog Heart

It is convenient to carry out this experiment on a perfused frog's heart set up as in paragraph 6.13. Two recording channels are required, one for heart movements, the other for the electrical activity. Use either a double-beam cathode-ray oscilloscope (preferably with storage facilities) or a two-channel ink-writer. The heart movements are transmitted to a transducer mounted over the heart in place of the usual lever. The electrical activity is picked up with saline-soaked wick electrodes applied gently to the surface of the heart. A second pair of wick electrodes are used to apply electrical stimuli during the cardiac cycle.

Explore the surface of the heart with the pick-up electrodes and relate the waveforms with the contraction of the atria and the ventricle. When you have identified the components of the cardiac electrogram, try the effects of stimulating the ventricle at various points of its cycle. Figure 6.9 illustrates the type of record obtained.

CIRCULATION IN MAMMALS

6.19 Recording of Blood Pressure in the Anaesthetized Animal

This procedure will be demonstrated; legal restrictions in Great Britain prevent the student from carrying out the experiment himself unless he holds a licence.

The arterial blood pressure is transmitted from a large artery through a fluid-filled tube to a manometer. In the past it was customary to use a mercury-filled manometer but this device has been rendered obsolete by transducer manometers of various types, see paragraph 1.7. The output from the arterial manometer is displayed by a recording galvanometer together with signals from other transducers attached to the animal.

The animal is weighed and injected subcutaneously or intraperitoneally: (a) cats, 45 mg pentobarbitone sodium (Nembutal) per kg., (b) rats and rabbits 1 mg. urethane per g. as a 25 per cent. w/v solution. Supplementary doses of anaesthetic are given as necessary, trichlorethylene vapour may be used to induce anaesthesia.

When the animal is fully anaesthetized an incision is made in the neck The trachea is exposed and a cannula inserted. The carotid sheaths are exposed on both sides and a length of carotid artery and vagus nerves separated by blunt dissection. A ligature is tied around one carotid artery as near the head as possible and a bulldog clip placed on the artery as far caudal as possible. The arterial cannula is inserted through an oblique incision and tied in position. The tube linking the cannula to the manometer is flushed free of bubbles with an anticoagulant solution, the system is closed, and the bulldog clip released. The recording galvanometer will now display the arterial pressure. So that drugs may be injected into the venous system, a cannula should be fitted to the external jugular vein.

The effects of a variety of procedures on the arterial blood pressure may now be demonstrated.

Stimulation of the Peripheral End of the Vagus Nerve. Record the arterial blood pressure and the electrocardiogram. Section the right vagus and lay the peripheral trunk on the stimulating electrodes. Apply a train of stimuli at 25 per sec, gradually increasing the voltage of the stimuli until an effect is obtained. Observe the effect on blood pressure, pulse rate and form of the electrocardiogram. Was the animal's heart rate changed by section of the vagus?

Injection of Drugs into the Circulation. Inject 25 μg of adrenaline dissolved in Locke's solution into the venous cannula. Observe the effect on heart rate, blood pressure and electrocardiogram. Try smaller and larger doses. When the pressure has returned to normal inject 1 μg of acetylcholine. Study the response. Now inject 1 mg atropine sulphate. Repeat the stimulation of the peripheral vagus; is the response the same as before? Repeat the injection of acetylcholine; has the response changed? Examine the effect of 0·1 mg. histamine phosphate.

6.20 Perfusion of the Mammalian Heart (Langendorff's Method)

The apparatus is shown in Figure 6.10. Oxygenated Locke's solution from a reservoir (R) is passed through a glass spiral contained within a bath of warmed water (B) to obtain an outflow temperature of about 38° C. The outflow tube has two side-arms, one of which accommodates the thermometer (T) while the other is connected to a mercury manometer (M) which indicates perfusion pressure. The outflow tube terminates in a cannula which is inserted into the aorta of an isolated heart (H). The size of the cannula depends on the size of the animal used.

Before beginning the experiment, see that the temperature of the Locke's solution issuing from the cannula is at least 36° C and not more than 39° C. Adjust the reducing valve on the oxygen cylinder so that a good stream of oxygen bubbles passes through the side-arm of the reservoir (R). Fill two evaporating basins with Locke's solution at room temperature for the reception of the heart. The animal is killed by a blow on the head and is bled out by cutting the carotid arteries. Quickly incise the skin in the mid-line over the sternum; then, holding the xiphisternum in forceps, make two cuts with scissors, one on either side of the sternum. Lift up the sternum and expose the heart by cutting through the pericardium. Hold the heart up with one hand and identify the aorta. Cut through the roots of the lungs, the venae cavae and aorta, leaving at least 1 cm of the aorta attached to the heart. Rapidly plunge the excised heart into the fluid in one of the evaporating basins so that it is completely immersed and keep pumping the heart with the fingers for one minute. In this way air cannot enter the coronary vessels, and blood clots are avoided. When the heart appears to be relatively free of blood, transfer it to the second clean basin. Carefully dissect away the remaining pericardium, identify the aorta and trim it if necessary. Make a short cut in the wall of the pulmonary artery to facilitate the passage of the outflow fluid. Loosen the clip above the cannula so that Locke's solution just drips out. Tie the first half of a ligature loosely around the cannula. With two pairs of forceps grip the wall of the aorta and slide it on to the cannula. Slip the ligature down, over the aorta, and tie it on to the cannula leaving as much of the blood-vessel as possible below the cannula. The flow of Locke's solution is now increased slowly by opening the screw clip. The Locke's solution cannot enter the left ventricle since the semi-lunar valves are shut by the pressure of the fluid, but it passes into the openings of the two coronary vessels and through the cardiac tissue to the venous side of the heart. Since the pulmonary artery is open, this fluid emerges at the base of the heart, flows over its surface keeping it moist and drips down from the apex. The first portion will be tinged red, but after five minutes the fluid emerging will be quite clear. The perfusion pressure should be fixed at between 50 and 70 mm Hg and the temperature in the cannula stablized at 37° C.

Now pass a threaded needle through the apex of the ventricles and tie the thread leaving two long ends. These ends should be attached to the rod on the right of the apparatus. With gentle tension, pull the heart slightly to one side. The fluid leaving the heart now follows the threads and is easily collected in a measuring cylinder to estimate coronary flow. Fasten a small clip (with thread attached) to the apex of the

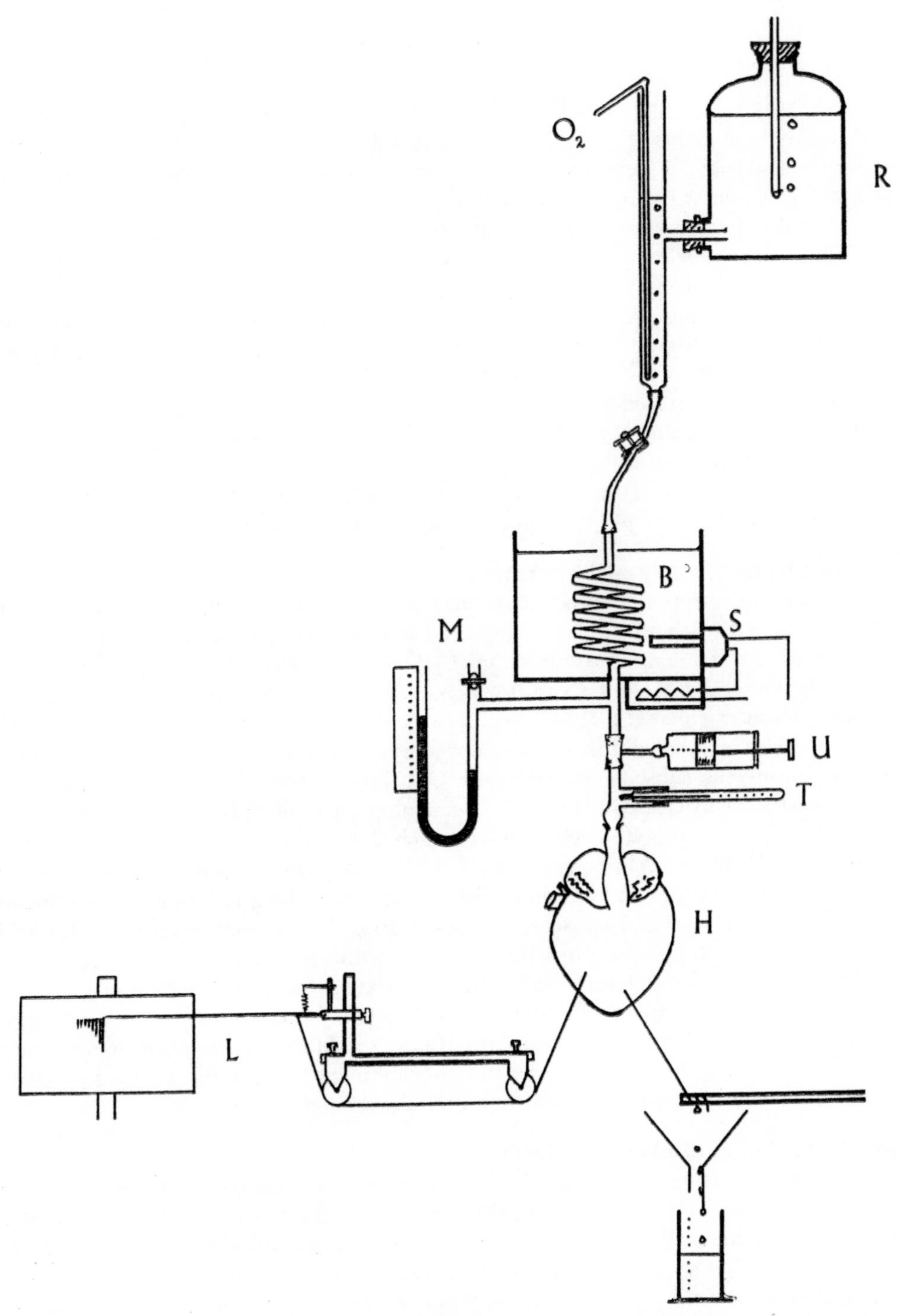

FIG. 6.10

Apparatus for the perfusion of the isolated mammalian heart as described in para. 2.19. R, Mariotte flask reservoir containing Locke's solution. B, Constant temperature bath with thermostat, S. M, Mercury manometer. U, Drugs are injected from a syringe whose needle is pushed through rubber tubing. T, Thermometer. H, Heart. L, Lever.

right ventricle and pass the thread over two pulleys to the heart lever. Record the amplitude and rate of beat on a smoked drum. The coronary flow is recorded as indicated above or with an outflow $recorder_1$.

After about 15 minutes the heart beats steadily. Measure the heart rate and coronary flow. Then introduce a small dose of adrenaline into the perfusion fluid by injecting it with a hypodermic syringe through the rubber tubing between the heart and the heating chamber. Note the increase in amplitude of beat and measure the heart rate and coronary flow. Repeat the procedure and observations with doses of acetylcholine, histamine and posterior pituitary extract.

Finally, apply electrodes to the surface of the heart and record electrograms.

This apparatus can be used to study hearts of rabbits, cats, guinea-pigs and rats. For rabbit hearts use the following doses: adrenaline 1 μg, acetylcholine 0·2 μg, histamine 25 μg and posterior pituitary extract 0·04 unit. For cat hearts use adrenaline 0·1 μg, acetylcholine 0·2 μg, histamine 25 μg and posterior pituitary extract 0·1 unit.

6.21 The Isolated Atrium Preparation

The isolated atria of a young rabbit may be used instead of the perfused heart for some experiments. A bath similar to that used in the study of rat diaphragm (Fig. 2.17) is employed at a temperature of 29 to 30° C. Oxygen is provided by a gas distribution tube at the bottom of the inner vessel which contains Locke's solution with double the normal amount of glucose.

The rabbit is killed and its heart removed as outlined in 6.20. It is kept in Locke's solution during the subsequent dissection. Cut through the atrio-ventricular junction to remove the ventricles. Cut away the remaining parts of the aorta and pulmonary artery. The parts of the atria which were attached to the aorta and pulmonary artery are now freed and are drawn gently apart. In this way the two atria remain in contact at the inter-atrial border and a strip of tissue about 5 cm long is formed. Tie a thread carefully on each end and suspend the preparation in the well-oxygenated Locke's solutions. Record the contractions on a smoked drum by means of a very light lever (such as a Starling heart lever). Leave the preparation for 15 to 30 minutes to settle down. Then find the effects of adding adrenaline (10^{-8}), acetylcholine (2×10^{-8}), and histamine (10^{-6}); wash out the bath with Locke's solution at the same temperature between each experiment. This preparation continues to beat for 12 to 24 hours, whereas the whole heart is unsatisfactory after about three hours.

6.22 Perfusion of the Isolated Rabbit Ear

In experiments on the intact animal, it is often impossible to say whether a fall of blood pressure is due to an action on the heart, or on the blood vessels, or both, and whether this action is direct or indirect (reflex or central stimulation). This preparation determines whether a substance has a direct action on the blood vessels.

The preparation is perfused at room temperature; the isolated hind legs of the rat, perfused through the aorta, can be similarly used, but the rabbit ear is the more sensitive preparation.

Preparation of Animal and Apparatus

The rabbit ear reaches its maximum sensitivity some 24 hours after its removal from the animal. For this reason, the dissection should be performed on the day preceding the one on which the experiment is planned.

You will be provided with a recently-killed rabbit. Cut out a strip of skin, about 1 cm wide, from the base of one ear so that the central artery comes into view. Separate a short length of the artery from the underlying tissue and tie a small cannula firmly into the artery. Cut off the ear. Attach a syringe filled with Locke's solution to the cannula and force the solution through the ear vessels, to wash out blood from the ear. Now connect the ear to a perfusion apparatus and allow Locke's solution to pass through the ear for a short time until no trace of blood appears in the effluent. The ear is now taken off the perfusion apparatus and placed in the refrigerator until the day of the experiment.

Experiment

Begin oxygenation of the perfusion fluid and connect the cannula, already inserted in the central artery, to the outflow tube from Mariotte bottle, A. Adjust the screw-clip until the Locke's solution is flowing steadily through the ear. Wait until the muslin ring on which the ear is placed is saturated and then choose a suitable time interval for interrupting the impulse-counter so that the resting flow rate gives a reasonably-sized trace on the smoked paper. Now try the effect of injecting the following substances, taking care to wait until the effect of one injection has passed off before making the next:

(1) 10 μg acetylcholine
(2) 0·01 μg adrenaline
(3) 0·05 μg noradrenaline
(4) 0·01 μg 5-hydroxytryptamine (Serotonin)
(5) Mephenesin.

Comment on your results and consider the difference in sensitivity between adrenaline and noradrenaline.

Now perfuse the ear with the fluid in bottle B, which contains benzylimidazoline (Priscol) 20 mg. per cent. Inject:

(6) 0·01 μg adrenaline
(7) 0·05 μg adrenaline
(8) 0·01 μg noradrenaline
(9) 0·05 μg noradrenaline.

Discuss these results.

Note: The doses given above are those which will be adequate for a very sensitive preparation; it may be necessary to increase them several-fold before definite effects can be observed.

CIRCULATION IN MAN

6.23 Heart Sounds

There will be a demonstration of heart sounds either pre-recorded (record or tape) or live, using a multi-channel stethoscope$_7$. It is easier to appreciate the general character and rhythm of the sounds in this way before attempting to use a stethoscope.

After the demonstration use the stethoscope; listen with the bell at the region of the apex beat in the fifth interspace about 9 cm (3½ in.) from the mid-line, at the lower left sternal border (tricuspid area) where the first sound is best heard and at the junction of the second right costal cartilage with the sternum, where the second (aortic) sound is clearest. The second pulmonic sound is most audible in the second left inter-space close to the sternum. In children and adolescents the second pulmonic sound is louder than the second aortic sound.

Note: Students intending to study clinical medicine are advised to purchase their own stethoscope at this time and should consult a member of staff about this.

6.24 The Pulse

The subject should sit at rest for five minutes so that any disturbance due to activity or emotion may pass off. Feel (palpate) the radial artery at the wrist with the tips of the fingers.

(1) *Rate:* Count the number of beats during one minute.

(2) *Rhythm:* Do the pulsations follow at regular intervals, i.e. is the rhythm regular? In some students it may be found that the heart rate increases slightly during inspiration. This is called sinus arrhythmia. Are all the pulsations of equal force?

(3) *Volume:* The volume or amplitude of expansion of the artery can be noted by light palpation; estimate it as large, medium or small.

(4) *Form:* Try to appreciate the normal rise, maintenance and fall of the wave in the artery. Compare the mental impression with the graphic record obtained later.

(5) *Pressure:* A very rough approximation to the systolic blood pressure may be obtained after some practice by digital compression of an artery in the following way. Three fingers are placed on the radial artery. Firm pressure is made with the distal finger to prevent any pulsation reaching the middle finger from the palmar arch. The proximal finger then gradually compresses the pulse till it can no longer be felt by the middle finger. This gives a very rough approximation to systolic pressure.

An idea of the diastolic pressure may be obtained by estimating the pressure required to flatten the vessel between beats. If the diastolic pressure is low, light compression will flatten the vessel, but considerable force will be required if the pressure is high.

It must be emphasized that although these two palpation methods can be, and often are, valuable, a quantitatively reliable estimate of the blood pressure can only be made with a sphygmomanometer (see 6.27).

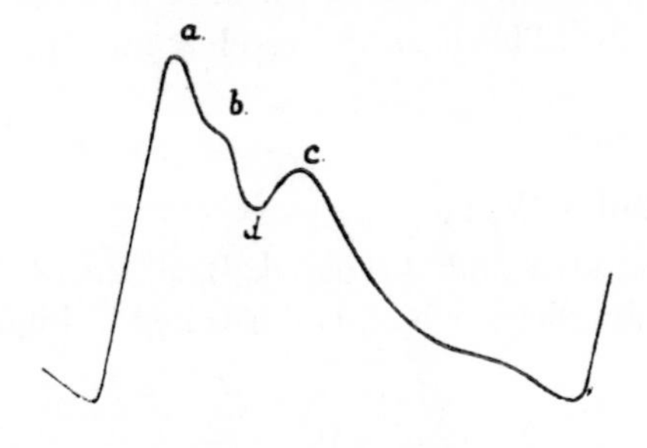

FIG. 6.11
a. Percussion wave
b. Tidal wave
c. Dicrotic wave
d. Aortic / Dicrotic } notch

(6) *State of the Vessel Wall:* Empty the artery by pressure with the finger and then palpate it longitudinally and laterally. The healthy vessel is impalpable or felt only with difficulty.

6.25 Recording of the Arterial Pulse-Wave

Two methods will be demonstrated and discussed. The records obtained from normal subjects all have the same basic shape which is illustrated in Figure 6.11.

(1) *Dudgeon's Sphygmograph*

This instrument is placed on the wrist so that the pulsations of the radial artery are transmitted to a small metal plate. Movement of the plate is amplified by a lever which writes on strips of smoked paper (15 cm × 2·5 cm) driven through the apparatus by a clockwork motor.

Smoke a kymograph drum and prepare strips using a guillotine provided with a check bar 2·5 cm from the edge of the base plate. It is important to cut the strips exactly to size and with parallel edges. Wind up the clockwork motor of the sphygmograph gently be means of the knurled screw; it can be started and stopped by the little lever on the top of the case. Insert a strip of smoked paper between the fluted roller (below) and the two wheels (above). Mark the position of the subject's radial artery with ink. Lay his forearm on a support a few centimetres above the bench level so that the wrist can be flexed or extended by him as required. Rest the spring button of the apparatus on the artery—keep the clockwork casing proximal. Do not strap the instrument to the wrist, but apply it with varying pressure in various positions of the wrist and at various settings of the eccentric pressing on the spring button till the maximum excursion of the writing point is obtained. Then allow the motor to drive the paper through the apparatus. Label the waves on the traces and make a note of the pulse rate on the tracing.

Now strap the sphygmograph on the wrist so that a good trace is obtained while the subject sits comfortably relaxed in a chair. The subject then stands up and another record is immediately made. Note the effect on the pulse of this sudden alteration of posture.

(2) *Photoplethysmographic Method*

The assembly, which contains a light source and small photoconductive cell is pressed against the pulp of the subject's thumb. Changes in blood vessel size with the arrival of the pulse wave cause changes in the optical properties of the tissue, and these are detected by the photoconductive cell, which is sensitive in the red and infra-red region of the spectrum. The change in the cell is amplified electrically and written out with a recording galvanometer. Note the resemblance to the pulse wave tracing produced by the Dudgeon's sphygmograph. Show the effect of an expiratory effort with a closed glottis (Valsalva's Manoeuvre).

6.26 Carotid Sinus Reflex

The subject must lie completely relaxed on his back. Feel for the pulsation of the common carotid artery deep to the anterior edge of the sternomastoid muscle. The bifurcation of the carotid artery and the carotid sinus are at the level of the upper border of the thyroid cartilage. Compress the artery firmly against the vertebral bodies for two seconds only. Note the slowing of the pulse. Other observers can feel the slowing at the radial arteries. Do **not** compress both carotids simultaneously.

Repeat this manoeuvre with the subject connected to the electrocardiograph (6.28).

6.27 Blood Pressure in Man

The sphygmomanometer consists of a valved pump and a rubber armlet which is connected to some kind of manometer. The best type of manometer is a glass U-tube containing mercury. If the two limbs are of the same diameter the scale is short (as in the original Riva-Rocci apparatus) but if one limb is made very wide the scale is much longer. Some patterns have an aneroid manometer which must be standardized against a mercury manometer. Preferably the subject should rest with his head and shoulders propped up at 45° during estimations but satisfactory results can be obtained if he sits quietly with muscles relaxed. It is essential that there should be no emotional or muscular disturbance because this may alter the readings considerably.

The subject's upper arm is bared to the shoulder. The centre of the completely deflated rubber bag is placed over the line of the brachial artery; the lower edge must be kept 2·5 cm above the bend of the elbow. Wrap the cloth bag which covers the rubber bag round and round the arm and tuck in (do not tie) the end under one of the turns. Position the manometer so that it can be seen by the observer but not by the subject. Make sure that the level of the mercury is at zero on the scale. The box must be on a level table and the scale should be read with the observer's eye at the same height as the mercury column.

Palpation Method

Palpate the subject's radial artery with the tips of the index and middle fingers of the left hand. Screw down the escape valve which is placed just above the bulb of the pump. Pump up the armlet rapidly (to about 150 mm) till the pulse disappears, then let out the air slowly by unscrewing the valve. Watch the manometer and note the reading when the pulse returns—this is the systolic blood pressure or S.B.P. Allow the pressure to fall rapidly once a reading is obtained.

Do not keep up the constriction any longer than is necessary; between readings let the pressure down to zero and allow the arm circulation to return to normal. This advice applies also to the next method.

Auscultation Method

This is the only reliable method for estimation of diastolic blood pressure (D.B.P.). It may be demonstrated by means of a microphone and loudspeaker. Examine the bend of the elbow when the subject pulls up his forearm against resistance. Note the tendon of the biceps and the sharp upper edge of the bicipital aponeurosis which runs medially from the tendon. You should be able to feel the pulsations of the brachial artery just medial to the biceps tendon. Hold the bell of the stethoscope with the left hand lightly over the site of maximal pulsation, but not in contact with the cuff, and inflate the armlet to about 30 mm Hg above S.B.P. as determined by palpation. If you are unable to feel the pulsations hold the bell just medial to the biceps tendon and proximal to the aponeurosis. Let the pressure down *slowly*. Above S.B.P. no sounds are heard. At S.B.P. successive tapping sounds are heard; note the reading on the manometer at their first appearance. This reading of the S.B.P. may be a few millimetres higher than that obtained by palpation. As the pressure is slowly reduced, the sounds increase to their maximum intensity and then decrease. They decrease gradually at first but suddenly change abruptly from loud tapping sounds to dull, muffled sounds or disappear completely. Where the sounds become muffled they, too, disappear with a slight further reduction in pressure. In Great Britain it is customary to take the point at which the loud tapping sounds change abruptly to the dull and muffled sounds as the diastolic pressure.

If palpation is used before the auscultatory method the unusual case with a silent gap will not be missed. In these cases the sounds are heard just below S.B.P. but they fade away and disappear only to reappear before D.B.P. is reached. The point at which the sounds reappear could be mistaken for the S.B.P. This silent gap is common in cases of aortic disease and high blood pressure. If the auscultation method is used exclusively, the mercury should be pumped to above 200 mm initially.

On listening with a stethoscope an attempt must be made to disregard all sounds other than those being investigated. Note that the taps are heard a fraction of a second before the mercury in the manometer bobs up a little. (The movement of the mercury is just a little late because of its inertia.) Sounds occurring between beats can be ignored. Record the readings and the pulse pressure, i.e. S.B.P. minus D.B.P.

6.28 Electrocardiography

The electrical potentials generated by the cardiac musculature can be detected at the body surface. The electrical disturbance recorded represents the algebraic sum of all the potential changes in the individual cells at successive points in time and the main deflection is of the order of 1 – 3 mV. A full explanation relating the electrical recording to the sequence of mechanical events in the heart is not yet possible and to introduce some uniformity in an otherwise empirical technique, certain set recording positions are used. These consist of:

1. Three bipolar leads which measure the potential difference between pairs of limbs,

Standard Lead I — Right arm and left arm (LA—RA)
Standard Lead II — Right arm and left leg (LL—RA)
Standard Lead III — Left arm and left leg (LL—LA).

2. The unipolar limb leads which record the potential difference between one limb and zero. The electrocardiograph obtains its zero potential for this purpose by summing the outputs of the RA, LA and LL and passing the resultant through a large resistance.

VR — Right arm
VL — Left arm
VF — Left foot.

Most electrocardiographs are constructed so that the voltages of the unipolar limb leads are amplified by an additional factor of 3/2 and the augmented records are prefixed by the letter A, i.e. AVR, AVL and AVF.

3. The unipolar praecordial lead, V, which records the P.D. between points on the chest wall and the same zero potential. A series of fixed positions on the chest wall have been selected and there is a V Lead corresponding to each. Seven of these are frequently employed.

V1. 1 cm from sternal edge in the 4th *right* intercostal space.
V2. 1 cm from sternal edge in the 4th *left* intercostal space.
V3. Halfway between V2 and V4.
V4. 5th left interspace in the mid-clavicular line (i.e. at apex).
V5. On the anterior axillary line, level with V4.
V6. On the mid-axillary line, level with V5.
V7. On the posterior axillary line, level with V6.

When a subject is connected to an electrocardiograph, not only are electrodes attached to RA, LA and LL, but a fourth is attached to the right leg (RL). This acts as an indifferent electrode and connects the subject to earth.

Each electrode consists of a metal plate (approx. 4·5 cm × 6 cm) held in position by a broad rubber band. To ensure good contact, a salt-enriched jelly is used between the electrode and the skin; the jelly usually contains powdered silica which acts as a mild abrasive, removing the very high resistance, outer layers of the epidermis. Electrically it makes no difference where the electrodes are placed on the limbs, except that a hairless area is preferable for best contact. This latter advantage is quite slight, however, and there is no need to employ depilatories. Routine use of the inner aspects of the forearms and of both legs just above the ankles gives satisfactory results. Two types of chest electrode are used frequently, both in conjunction with jelly. The first is a small metal disc (1 cm diameter) mounted on a short bakelite rod which must be held by the subject; the other incorporates a sucker to hold it in position.

Examine the electrocardiograph provided; a number of models are available and the features of the model used in this laboratory will be demonstrated. They all share certain principal features.

1. A 'Lead Selector' switch which is marked in a standard order, namely, Standardize, I, II, III, AVR, AVL, AVF and V (any additional Leads can be ignored at this stage).

2. A 'Balance' control with which unwanted electrical potentials can be attenuated.
3. A trigger, also marked 'Standardize' which allows 1 mV pulses to be fed into the circuits.
4. A 'Gain' control with which the deflection produced by 1 mV is adjusted to 1 cm (occasionally marked 'Sensitivity').

Modern instruments usually record by heated stylus on heat-sensitive paper. A paper-drive speed of 2½ cm/sec is used in clinical practice.

Connect a subject to the electrocardiograph and secure sample records of the leads listed above. To avoid subsequent confusion, label each as it is recorded. Taking an ECG is easy; interpreting it is the difficult part. A full account cannot be given here but a routine procedure should be followed which includes the following points at least. In clinical terminology the word 'lead' is applied to the actual paper record.

1. Determine which leads have been taken and whether they are correctly labelled.
2. Check the calibration, which should always be recorded; i.e. that the standardizing 1 mV produced a deflection of 1 cm.
3. Note any artifacts present; oscillations due to contractions of skeletal muscle and interfering extraneous voltages are the commonest artifacts.
4. Determine the fundamental ventricular rhythm and rate, and the relationship between atrial and ventricular rhythms. Name any arrhythmia present.
5. Note the form, amplitude and duration of the P wave.
6. Measure the P-R interval.
7. Note the amplitude, form, electrical axis and duration of the QRS complexes.
8. Note the form and any displacement of the ST segment.
9. Note the form, amplitude and electrical axis of the T waves.
10. Measure the duration of systole (Q-T).
11. Note the relationship between the electrical axes of QRS and T.

After sample tracings of the Leads have been obtained, turn the Lead Selector Switch to STANDARD LEAD I and perform the following experiments.

1. Stimulation of the carotid sinus (6.26).
2. Effect of posture on pulse rate (6.30).

Do a beat by beat analysis of the records you obtain. The distance between successive beats is the reciprocal of rate. Display your results graphically with Distance (between beats) on the ordinate and Successive Beats along the abscissa, indicating where each event, e.g. carotid compression, took place.

6.29 Simultaneous Recording of Electrocardiogram, Heart Sounds and Carotid Pulse

A three-channel inkwriter is used. The electrocardiogram (lead 2) is fed into channel 1. The output from a contact chest microphone is fed into channel 2 and the output from a pressure transducer applied to the skin of the neck over the carotid artery is fed into channel 3. The subject lies supine on a couch or bench-top and the ECG limb electrodes are applied. He bares his chest and the standard positions for heart sound observation are noted, as described in 6.23. The chest microphone is supported by light beam spanning the couch and its position adjusted by a vertical rackwork to press on the subject's chest. The carotid pulse transducer is applied over the region of maximum pulsation in the neck at the level of the thyroid cartilage. When the gain on each channel has been adjusted to give a suitable excursion on each pen, the subject is told to stop breathing and the paper feed mechanism on the recorder is started. Record at 2·5 cm/sec for about ten seconds, stop paper feed and tell subject to restart breathing. Apnoea during recording reduces noises which would otherwise partially obscure the sound record. From your record (an example is shown in Fig. 6.12), consider and relate the cardiovascular events and give an explanation. The trace from the chest microphone is called a phonocardiogram.

EFFECT OF CHANGE OF POSTURE ON CARDIOVASCULAR SYSTEM

6.30 Pulse

The subject lies horizontally for a five-minute rest period. Then count his pulse for 30 seconds every minute for five minutes. He now stands up and his pulse is counted in consecutive 30-second intervals for five minutes.

He now lies down and his pulse is again counted over 30-second periods.

Prepare a graph of these results. What was the percentage rise in pulse rate on moving from the horizontal to the vertical position?

6.31 Blood Pressure

Repeat experiment 6.30 but record systolic and diastolic blood pressure at minute intervals instead of pulse-counting. Graph these results.

6.32 The mechanisms producing the changes noted in 6.30 and 6.31 can be further investigated by strapping the subject to a tilt-table. Changes in posture can then be effected without muscular effort on the subject's part and, if he is suitably phlegmatic and conditioned to the experience, without any additional mental activity or anxiety.

The subject mounts the tilt-table which is swung into the horizontal position. He adjusts his position till the table is in balance, the foot-rest is adjusted and he is strapped into the harness.

Repeat experiments 6.30 and 6.31. Write a short note on the possible mechanisms underlying the cardiovascular responses you have demonstrated. What conclusions can be drawn specifically from the use of the tilt-table?

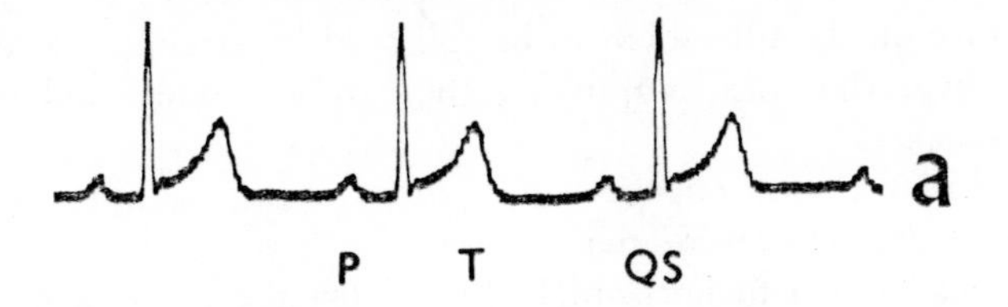

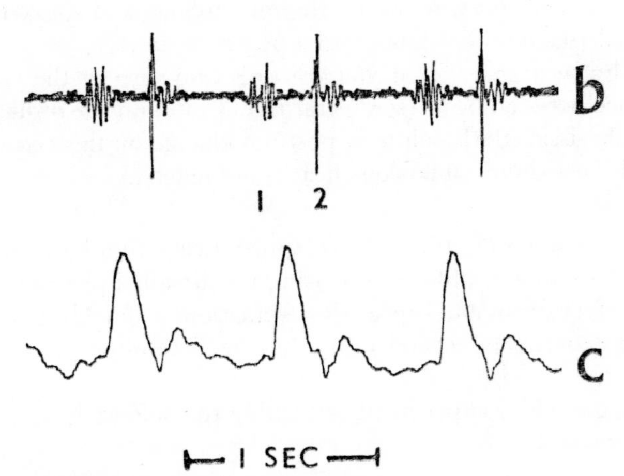

Fig. 6.12

Record from a three-channel inkwriter. a, is lead 2 from the electrocardiogram; b, heart sounds (1 and 2) from a chest microphone; c, the carotid arterial pulse. Note the time relationships of the wave forms.

6.33 The rapidity of, and to some extent the magnitude of, pulse-rate changes with changes of posture on the tilt-table can be followed by strapping a microphone to the subject's chest, over the apex, amplifying the cardiac sounds and reproducing them through a loudspeaker.

Note that the maximum change in frequency occurs within a few seconds of the change in posture. Perform several manoeuvres, e.g. head down to feet down, feet down to head down, head down to horizontal. Try to list the movements in magnitude of effect produced on the pulse rate. How does rate of tilting affect the response?

6.34 Repeat experiment 6.33, replacing the microphone and speaker with the electrocardiographic leads feeding the signal into a pen-recorder.

Make beat by beat analyses of your records and display the results in graphical form with distance between beats (the reciprocal of rate) on the ordinate and successive beats along the abscissa. Mark points of postural change on this record.

Alternatively, use an instantaneous heart rate meter[11].

6.35 Place the subject on the tilt-table as before. Leave him horizontal for 15 minutes and then estimate his forearm blood flow using the air-filled plethysmograph (see 6.44). Now tilt him 45° feet down and repeat the estimation as quickly as possible.

Perform the experiment several times. Try to determine the minimum tilt which produces an effect.

Then repeat the whole experiment but tilting the subject head down 30°.

Discuss your results.

6.36 Effect of Vasodilatation on Blood Pressure

Anyone who has had any 'heart trouble' must not act as subject in this experiment. Make a table with three columns for Time, S.B.P. and D.B.P. Record normal blood pressure over three consecutive minutes. Then ask the subject to inhale deeply through a piece of cotton-wool on which two drops of amyl nitrite have been placed. Record the S.B.P. and D.B.P. at minute intervals (or every two minutes if this is too difficult) and continue to take readings for several minutes after normal B.P. has been regained. While the observations are being made a second observer should count the pulse rate for half a minute at minute intervals. Watch also the skin colour and respiration of the subject. When the experiment is over ask the subject to describe his sensations after inhalation of the drug.

Draw a graph: abscissa 2·5 cm=5 min; ordinate 2·5 cm=20 mm Hg.

6.37 The Venous Flow

Apply a sphygmomanometer cuff just above the elbow and inflate to 40 mm Hg. Observe the appearance of the veins in the forearm. Note the occurrence of little swellings on the course of the veins, showing the position of the valves.

Place a finger (A) on one of the veins and note the position of the valve (V) next above it. With another finger press the blood in the vein towards the upper arm keeping finger (A) in the same position. There will be no influx of blood from above and the portion of the vein between (A) and the valve (V) will remain collapsed. The vein above (V) will be distended and the valve will show clearly. Still keeping finger (A) in the same position place another finger above the valve (V) and attempt to press the blood towards the wrist. The blood cannot be forced through unless pressure sufficient to rupture the valve is used. Remove finger (A) and note that the vein which was collapsed is immediately filled from below. These are some of Harvey's original experiments. Can you recall any other evidence for the circulation of the blood? What is the function of the valves in the veins?

6.38 The Jugular Venous Pulse

The subject lies supine on a couch with a small pillow under his head. His chin should point slightly to the left. Examine the surface of the right side of his neck under good illumination and identify the sterno-mastoid muscle, the trachea and the thyroid cartilage. Ask the subject to make an expiratory effort against a closed glottis (the Valsalva manoeuvre) and notice how the superficial veins fill and swell with blood. This occurs because the entry of venous blood into the thorax has been impeded by the positive pressure produced by the expiratory effort. Identify and mark with a grease pencil the large superficial vein which runs over the sternomastoid muscle close to, and roughly parallel to, its lateral margin. This is the external jugular vein. During quiet breathing it should show pulsations which vary with respiration; if the pulsations are not obvious, vary the height of the pillow supporting the head. The apparatus provided is designed to pick up and record the pulsations of the vein and relates these to the heart's action. This can be done by recording the subject's electrocardiogram simultaneously with the jugular pulse and displaying the record on a two-channel inkwriter or on a double-beam cathode-ray tube with a long-persistence screen (or storage oscilloscope). The jugular pulse is recorded by pressing a tambour with a slack rubber diaphragm over the region of maximum pulsation: the pressure waves from the tambour are conducted along a short rubber tube to a high sensitivity pressure transducer. The electrical ouput feeds the display device. It is convenient to have a side-tube together with clip in the tube between tambour and transducer. The clip is opened between recordings. The pulsations vary with respiration and the subject should be asked to stop breathing in the expiratory position while recordings are made. A record is shown in Figure 6.13; note the three waves marked A, C and V and their relationship to the P, QRS and T waves of the electrocardiogram. Read up the theory of their formation in a text-book.

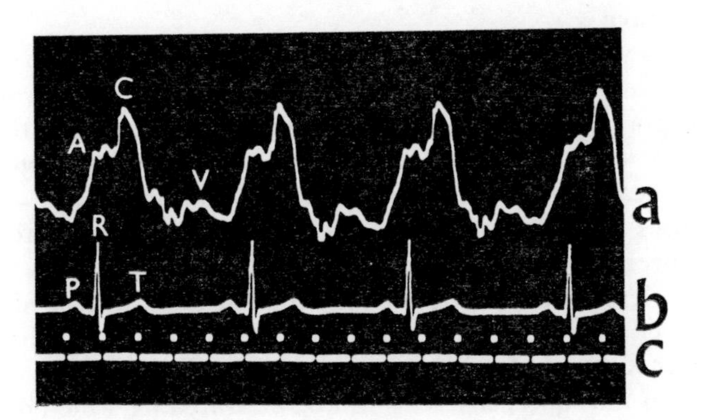

FIG. 6.13

Photograph from a cathode-ray tube face. a, the jugular venous pulse, showing A, C and V waves, detected by a pressure transducer. b, the electrocardiogram, lead 2; P, R and T waves are labelled. c, time trace 0·2 sec.

6.39 Venous Pressure

The subject sits with his arm hanging vertically downwards. The veins of the arm will become distended. The observer slowly raises the subject's arm till the veins above the wrist begin to empty. Measure the vertical height between the wrist and the third costal cartilage at its sternal end, i.e. the level of entry of the superior vena cava. This distance is a measure of the pressure in the right atrium.

A vein collapses if the external pressure, i.e. atmospheric pressure is greater than the internal pressure; it remains full if the internal pressure exceeds the atmospheric pressure. Thus the level at the junction of the collapsed and distended portions of the vein is at atmospheric pressure. If the atrium is below this level, the atrial pressure can be expressed as X cm of blood above atmospheric.

The pressure in the superficial veins can be measured by a method similar to that used for determining arterial pressure. Stick a flat glass cup to the skin of the forearm over a vein by means of collodion. When the collodion is dry connect the outlet tube from the cup to a water manometer and rubber bellows. Increase the pressure in the system till the vein is just made to empty; the reading of the water manometer at this point gives the pressure in the vein. Assuming that there is a continuous blood column up to the heart from the part of the vein covered by the cup, again calculate the pressure in the atrium at the point of entry of the superior vena cava.

With the subject in the supine position observe the pulsation of the external jugular vein. What is the effect on the pulsation of tilting the subject a little, first head up and then head down?

6.40 Response of Skin Capillaries to Injury

Draw a blunt-edged instrument, e.g. a closed forceps, firmly across the anterior surface of the forearm. The reaction varies with the amount of the injury done to the skin. Moderate pressure will only displace the blood from the skin locally. The reaction varies from person to person, but when fully developed it consists of:

1. A red line (tache) in the track of the instrument.
2. A surrounding ill-defined scarlet area or flare.
3. A local oedema or even a weal developing under the track.

Lewis suggested that a histamine-like substance is liberated in the skin on injury and is the cause of this triple response.

Compare the effects noted above with the reaction of the skin capillaries to histamine and adrenaline. Wash the forearm with soap and water, dry, and clean with spirit on a piece of cotton-wool. On the skin place drops, 2·5 cm apart, in this order from above downwards:

1. Locke's solution. 2. 1:10,000 histamine. 3. 1:1000 adrenaline.

Flame a needle and scratch through 1 firmly but do not draw blood. Flame the needle again and scratch through 2. Flame and scratch through 3.

Describe the appearances in 2 and compare them with the triple response.

Explain the occurrence of the oedema.

Describe the reaction to adrenaline. Look for cutis anserina (goose skin) and explain the appearances, related if possible to the nerve supply of the structures involved. Place a sphygmomanometer cuff on the arm and blow it up to 40 mm Hg. to raise the venous pressure in the limb. Does the blanching disappear? If not raise the cuff pressure in steps of 10 mm Hg. and report the results.

6.41 Bleeding Time

(1) *Duke's Method.* Wipe the lobe of the subject's ear gently with a spirit swab and allow the skin to dry. Sterilize a straight needle by heating it in a flame or use a disposable sterile lancet[3] and make a single puncture 4 mm deep into the lobe of the ear. A spring-loaded stylet which can be set to any chosen depth helps to standardize the procedure. Thirty seconds later remove the drop of blood by applying a clean piece of filter paper. Repeat the procedure every 15 seconds using a fresh area of the paper on each occasion until bleeding ceases. Count the spots of blood. The normal bleeding time by this method is from two to seven minutes. Collect results from at least a dozen subjects to get an impression of the normal scatter.

(2) *Ivy's Method.* Clean the anterior surface of the forearm with spirit. Put a sphygmomanometer cuff round the arm; raise the pressure to 40 mm Hg. and maintain this pressure till the end of the experiment. Use a sterile disposable lancet and make a puncture 2·5 mm deep into the skin on the anterior surface of the forearm. Remove the drop of blood on filter paper as in the previous method. The bleeding time by this method is rarely more than 180 seconds, but up to 240 seconds is considered normal. Compare the results obtained from other members of the class.

6.42 Observation of Skin Capillaries

This should be performed on your own finger. Place the hand palm downwards on the bench. Adjust the illuminator, which has a heat filter, so that a brilliant spot is focused on the cuticle of one nail. Apply a drop of liquid paraffin or immersion oil. Focus the binocular stereoscopic microscope (×20) on capillary loops.

The capillary loops of the skin usually lie at right angles to the surface so that only a part of each loop is in focus at any one time. If however, the cuticle (eponychium) has been left undisturbed for at least a week and is allowed to grow over the nail, the capillary loops will be found lying obliquely and very often a whole loop can be seen in good focus at one time. Adjust the illuminator so that reflections from the skin surface cease to be troublesome.

Select one loop for prolonged observation. Does its diameter remain constant? Does it ever shut down completely? Can individual red cells be seen? If so, describe their movements. Place a sphygmomanometer cuff round your arm and raise the pressure to 50 mm Hg. to occlude the veins. Describe the appearance of the capillaries. Release the pressure for a little and then put it up to a value midway between systolic and diastolic blood pressure (say 100 mm Hg.) Report on the behaviour of the capillaries. Finally let the pressure fall to zero for a few minutes and then raise it suddenly to 150 mm Hg. to occlude the arteries. Describe what happens and note the time taken for the flow to come to a standstill. Release the pressure and remove the cuff.

Pull the end of a solid glass rod out to a fine point in a Bunsen flame. Break off the end, leaving a few centimetres of fine glass needle attached to the rod which acts as a handle. Push it under the cuticle, and under microscopic observation push it on till it reaches the convexity of a capillary loop. When the capillary is in good focus push the needle into it. Describe what happens to the capillary.

6.43 Digital Plethysmograph

The finger plethysmograph is a small glass container (Fig. 6.14) which is sealed to the finger with petroleum jelly. A length of rubber tube connects it to a horizontal glass tube 25 cm long, calibrated in 0·01 ml., containing a short column, say 1 cm in length, of coloured alcohol the movement of which indicates changes in the volume of the finger. The bulbs at either end of the calibrated tube prevent loss of the alcohol.

The subject should be warm, or the skin vessels will be constricted and an unsatisfactory result obtained. Insert the middle finger in the plethysmograph to the proximal interphalangeal joint, and seal off the space round the finger with petroleum jelly. With the finger in position the alcohol column is set to lie near the middle of the calibrated tube. This is done by adjusting with a screw clip the two metal plates on the rubber side tube.

(1). Note that small movements of the hand cause large movements of the alcohol column. Try to keep the hand as still as possible by assuming a comfortable relaxed position.

(2). Watch the pulse waves. Measure their amplitude in 0·01 ml.

(3). Inflate a sphygmomanometer cuff round the upper arm to 200 mm Hg. for five minutes to arrest the circulation. Any movements of the alcohol column must now be due to involuntary movements of the finger in and out of the plethysmograph. Note the extent of the movement which can be ascribed to this cause. Near the end of the five-minute period adjust the position of the alcohol so that it is about 3 cm from the plethysmograph end of the tube. On release of the cuff note the amplitude of the pulsations at each heart beat. Observe also the slower movement of the alcohol column from its initial position. During the period of reactive hyperaemia the finger swells because the vessels dilate and contain more blood; the pulsations are increased in amplitude. Continue your observations till the reactive hyperaemia has passed off.

(4). Adjust the alcohol column to the middle of the tube. Note that there are fluctuations in finger volume of an irregular character. If these are greater than those seen in (3) after occlusion of the circulation it suggests that the finger vessels are undergoing irregular rhythmic changes in calibre. This is in fact the case. It is found that these fluctuations occur simultaneously in all extremities and are the result of variations in the discharge pattern of the sympathetic nervous system. The amplitude of the pulse wave may also vary from time to time.

(5). With the least possible movement of the hand take a deep breath. Note that the finger volume decreases; this diminution is noticeable within a few seconds and reaches its maximum in approximately ten seconds. This result is due to a reflex constriction of the skin vessels, but neither the origin of this reflex nor its purpose is understood. The vessels constrict in response to impulses travelling in the sympathetic nerves.

(6). Arrange a sphygmomanometer cuff round the wrist. Adjust the alcohol so that it is near the plethysmograph end of the tube, and observe its initial position for about a minute or until it is reasonably steady. Inflate the cuff to 70 mm Hg. and watch the alcohol. It moves across at first with a uniform speed and then more slowly. The cuff prevents, for a time, the return of blood from the hand to the heart without interfering with the arterial inflow. Thus the hand and the finger swell at a rate equal to the rate of entry of blood in the arteries. The rate at which the alcohol moves initially is thus a measure of the rate of blood flow into the finger. This does not give an absolute measure of finger blood flow because some of the blood passing to the finger returns and accumulates in the body of the hand. For accurate measurement of finger blood flow a cuff is inflated round the base of the finger instead of round the wrist. Movements of the alcohol may be recorded photographically.

(7). Set up the plethysmograph as described using the right middle finger and, when the subject is relaxed, place a sphygmomanometer cuff around the left arm and inflate it to 200 mm Hg. Immerse the left forearm in hot water and observe the rise in volume of the finger in the plethysmograph. When the reading has become steady allow the air to escape from the cuff and observe a further rise in the volume of the finger in the plethysmograph. Discuss the vasomotor behaviour of the finger in the plethysmograph as a reflex phenomenon and as a response to changes in blood temperature.

Measurement of Arm Blood Flow

The measurement of blood flow is especially difficult in man since direct interference with the vessels is not usually permissible and circulatory disturbances are very readily produced. Skin temperature readings reflect changes in blood flow but do not give quantitative measurements.

Plethysmographic technique is applicable to certain regions of the body in both man and animals. It depends on the fact that in any organ if the venous return is suddenly obstructed, the organ increases in size. The increase depends on the structure of the organ to some extent, and can, of course, only continue for a short time until the organ becomes congested with blood. During the period immediately following venous occlusion the increase in size is due entirely to the entrance of arterial blood. The plethysmograph consists of a closed rigid cylinder of metal or glass placed around the organ. It communicates with a volume recorder.

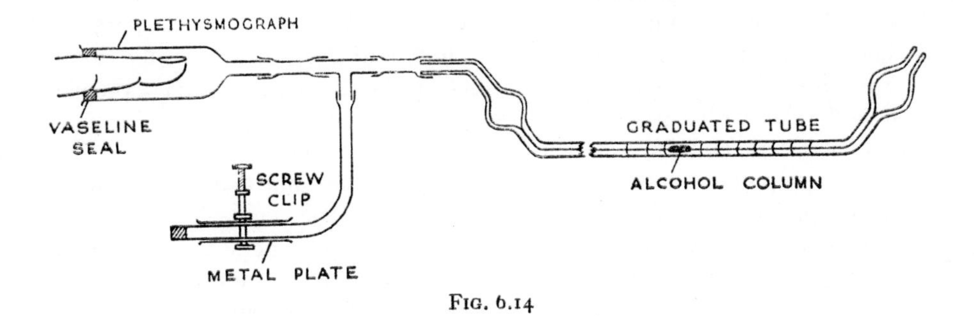
PLETHYSMOGRAPH
VASELINE
SEAL
SCREW
CLIP
METAL PLATE
GRADUATED TUBE
ALCOHOL COLUMN

FIG. 6.14

6.44 Simple Air-filled Plethysmograph

Open spring clip A (Fig. 6.15) and place the arm in the plethysmograph. The plethysmograph must be mounted sufficiently high with respect to the subject's thorax for the arm veins to be collapsed, see paragraph 6.39. Mark the upper level of the plethysmograph on the skin with a grease pencil then make an air-tight junction at the upper end by means of an elastic cuff. Be sure that this junction is air-tight but make the tension of the cuff as low as possible so that the venous return is not impeded. A sphygmomanometer cuff is applied above the elbow ready to constrict the veins when required. The volume recorder can consist of a small counterweighted bell or float, like a miniature spirometer with the same sort of water seal, or alternatively a piston recorder$_1$. In each case it must be counterbalanced and move freely. Position is transmitted by a pointer which records on a smoked drum. If a piston recorder is used it must be kept dry and free from dust. When the limb in the plethysmograph increases in volume as a result of venous occlusion, air is displaced into the recorder and the pointer rises. The whole recording apparatus is delicate and requires careful attention.

Keep spring clip A open until you are ready to take a measurement, for both volume recorders are sensitive and movement of the arm in the plethysmograph may suck water into the rubber tubing of the first, or damage the piston of the second. Once the arm is in the plethysmograph, spring clip B may be opened and A shut; after this the subject must remain perfectly still and relaxed, leaving all adjustments of the apparatus to his partner.

The drum should revolve slowly, about 1 cm in five seconds, with the pointer of the volume recorder adjusted to write lightly and clearly. If the cuff is airtight and the float balanced, small waves due to the pulse and larger waves due to respiration will appear on the smoked surface. If these small oscillations are not obtained the apparatus is not adjusted correctly. Do not proceed further until this degree of sensitivity is achieved.

Now set the pointer to its lowest position by opening clip A. Sometimes the subject by a *very slight* withdrawal of the arm can adjust the pointer, but this must be done with care. Start the kymograph; after a length of level trace is obtained suddenly inflate the sphygmomanometer cuff to 70 mm Hg. and keep it there until the trace ceases to rise. Release the air pressure in the cuff, stop the drum and disconnect the plethysmograph by closing clip B.

If a water-filled volume recorder is used it is important not to make any attempt to readjust the float, the cuff, or the position of the arm until the plethysmograph has been disconnected from the volume recorder. If a mistake is made in this respect water will either be blown out of the system or sucked in; this does no harm but necessitates clearing the tubing out before proceeding.

From the tracing obtained in this experiment the arm blood flow at rest can be calculated. Run a time-trace on the drum; either one-second or five-second interval marks are suitable. Calibrate the volume recorder before the paper is taken off the drum. Two methods are available. For piston recorders which must be kept dry inject

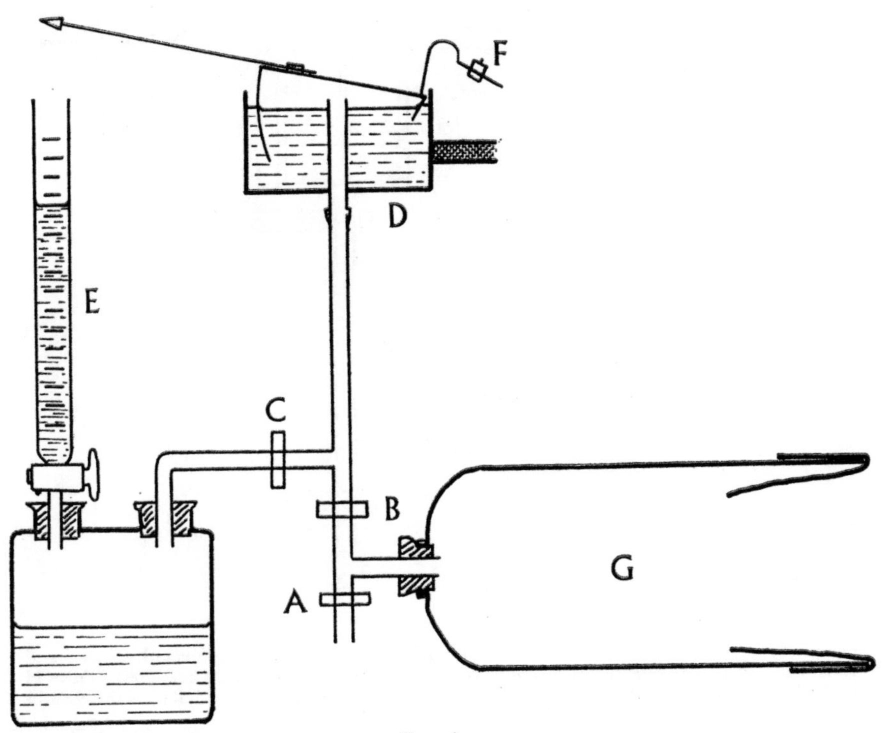

Fig. 6.15

A simple air-filled plethysmograph for forearm and hand. The plethysmograph, G, with a rubber cuff to seal the junction with the forearm, is connected to a volume recorder, D. In this case a float recorder is shown, but a piston recorder as shown in 6.16, may be used. Counterweight F is used to balance the float. The side tube connects with a burette E and Woulfe's bottle for calibration of the system. Clips A, B and C are referred to in para. 6.44.

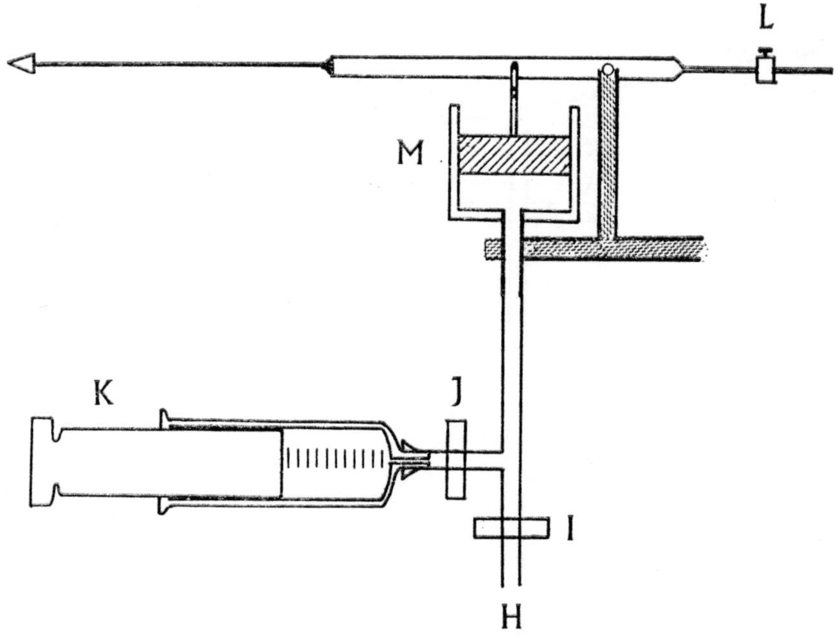

Fig. 6.16

Piston recorder, M, fitted with side tube to which a syringe, K, is connected. Clips J and I are used to isolate the recorder from syringe or plethysmograph as required. Tube H is connected to the plethysmograph. Counterweight L is adjusted to balance the weight of the piston.

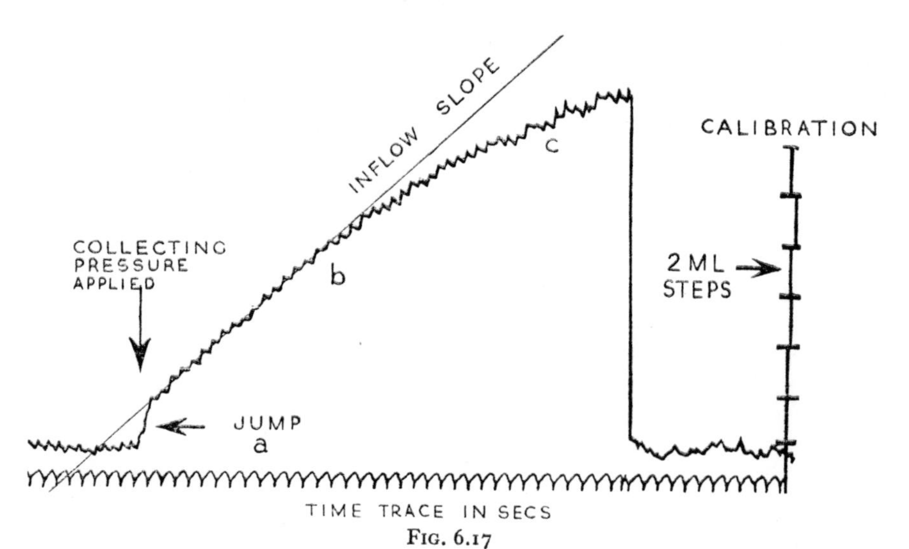

Fig. 6.17

An example of the tracing obtained by venous occlusion plethysmography.

known volumes of air into the system with a syringe after closing spring clip I (Fig. 6.16). For float recorders the burette of Figure 6.15 is used. Empty the Woulfe's bottle and insert the burette into one neck. Fill the burette with water. Connect the other neck of the bottle to the volume recorder through the spring clip C as shown in the diagram so that the bottle replaces the arm and the plethysmograph. With each method set the poiner horizontal, and move the drum round a short way by hand to record the zero. Now run in 2 ml. of water or air; mark another line here and at each 2 ml. interval thereafter until the recorder reaches the upper limit of its excursion.

Measure the volume of the arm by inserting it into a glass cylinder. Fill the cylinder to the brim with water. Insert the arm gently to the level which was marked when the estimations were begun. Withdraw the arm and measure the volume of water required to refill the cylinder. This is the volume of the forearm.

The analysis of the records is best done after the tracing is varnished and dried. The record will have the general appearance shown in Figure 6.17. It shows (a) an immediate jump or artefact due to movement of the tissues at the moment of inflation (the arm and the apparatus should be arranged so that it is as small as possible); (b) a steadily climbing line which is used to calculate the rate of inflow; (c) a line with diminishing slope. As the arm vessels become distended and the venous pressure rises above the cuff pressure, blood escapes back under the cuff towards the heart and the volume of the arm approaches a maximum value. In addition the increase in venous pressure opposes and slows up arterial inflow.

Draw a sloping line through the initial portion of the curve as indicated on Figure 6.17. From the calibration curve and the time trace estimate the blood flow in ml. per minute. Divide this figure by the volume of the arm in ml. and multiply by 100 to obtain blood flow in ml./100 ml. arm/min.

6.45 Reactive Hyperaemia

Put the arm in the plethysmograph but do not connect it up to the volume recorder. Inflate the sphygmomanometer cuff round the arm to 200 mm Hg. and keep it at this pressure for five minutes by repeated pumping as necessary. At the end of the time connect the plethysmograph to the volume recorder, start the drum moving and suddenly drop the cuff pressure to 70 mm Hg. Neglect the initial disturbance arising from muscle movement and note the greatly increased blood flow.

6.46 Effect of Work

In another subject measure the resting flow as before. While the arm is in the plethysmograph the subject clenches his fist hard 25 times in rapid succession. Connect the plethysmograph to the volume recorder and take another trace.

6.47 Effect of Heat and Cold

Hold the forearm in water at 44° C for five minutes or more then insert it into the plethysmograph and measure the blood flow. In the same way measure the blood flow after the temperature of the forearm has been considerably reduced by holding it in cold water for a few minutes.

6.48 Blood Flow in the Human Forearm

The forearm is composed mainly of muscle and bone, with relatively little skin. It is chiefly the blood flow through the muscles which determines the total forearm blood flow. Total flow may be measured by enclosing the forearm in a plethysmograph$_{33}$ and recording changes in its volume. When a cuff is applied just above the elbow at a pressure of 70 mm Hg. the entry of arterial blood into the forearm is unimpaired, but for some time venous blood cannot return to the heart. The blood collected in the forearm causes it to swell and the increase in volume is recorded. The method given in this section incorporates several refinements not included in the simple apparatus described before (para. 6.44).

There is a flange at each end of the plethysmograph, to which a rubber diaphragm is bolted (Fig. 6.18). A sleeve of light rubber dam placed inside the plethysmograph is attached at either end to openings in the diaphragms. The hand and forearm are powdered and inserted through the openings in the diaphragms so that the upper end of the plethysmograph is close to the elbow. This ensures that as much of the forearm musculature as possible is included. There are three openings in the upper surface of the the plethysmograph, one for a thermometer, one for a stirrer and a large one, 2·5 cm in diameter, surmounted by a vertical glass cylinder. Through the latter the plethysmograph is filled with water at 34° C until the water rises 3 to 5 cm in the glass cylinder. The pressure of the water keeps the rubber dam in contact with the forearm.

Wrap the 4 cm broad pneumatic cuff round the wrist and connect to a sphygmomanometer so that when necessary it can be inflated to 200 mm Hg. to arrest the circulation to the hand. Apply the 'collecting cuff', which is 10 cm broad, just above the elbow. This is connected by a rubber tube and a glass tap to an air reservoir and a mercury manometer. The pressure is maintained at 70 mm Hg. in the reservoir by means of a foot-pump or a sphygmomanometer bulb. When the tap is opened the collecting cuff is suddenly inflated to 70 mm Hg. and blood collects in the forearm, chiefly in the veins. If the veins are distended at rest the method becomes inaccurate and the subject should, therefore, be seated with his forearm just above heart level.

A syringe, as shown in Figure 6.16, is connected by a side tube to the pipe linking the plethysmograph to the recorder; it serves to adjust the level of the recorder (either piston as in Fig. 6.16 or float as in Fig. 6.15) or to calibrate the plethysmograph assembly, by injecting known volumes of air into the system. If the air is injected in steps of two ml., a slight backward and forward movement of the kymograph at the end of each step produces a scale on the kymograph paper (Fig. 6.17). When the volume of the forearm increases the same amount of water is displaced from the plethysmograph and this in turn displaces air through the rubber tubing to the recorder which writes on smoked paper. The clamp holding the recorder should be adjusted so that the pointer moves in a vertical plane at right angles to the base-line on the kymograph.

Adjust the speed of the kymograph to about 15 cm per minute. Make a record with the arm at rest and without any pressure in the cuffs. At the same time make a one-second time-trace. The recorder will give a level trace, with small increases of

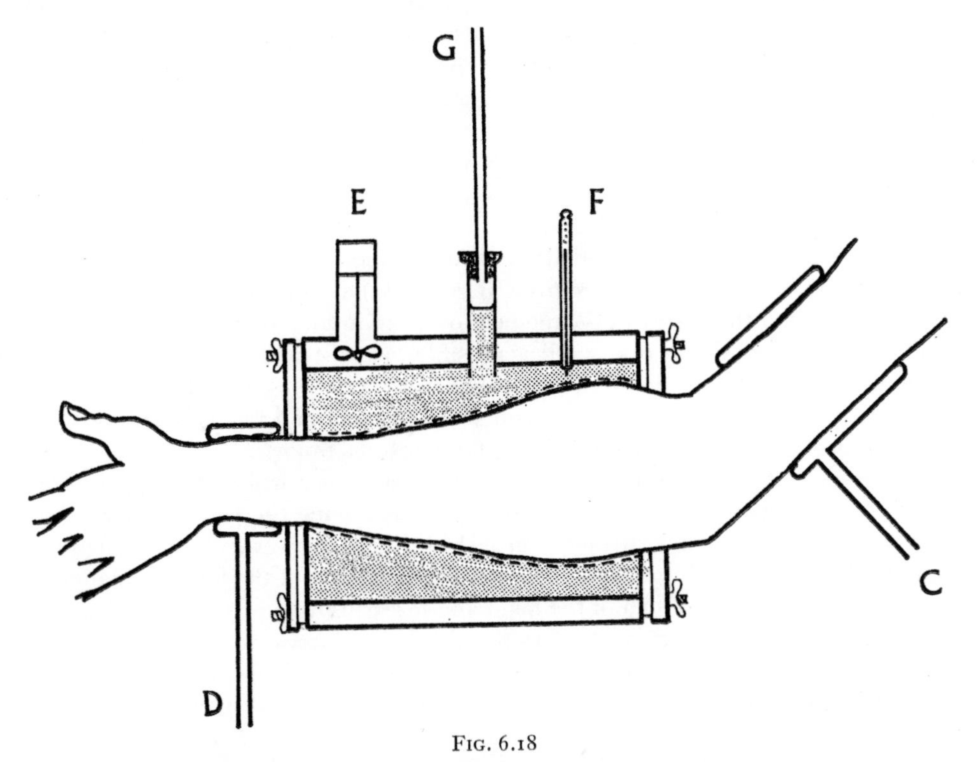

FIG. 6.18

Water-filled, temperature-stabilized, forearm plethysmograph.
The procedure is described in para. 6.48: air from the reservoir is led into the collecting cuff by tube C. The wrist cuff is inflated via tube D. The outer water-jacket, stirred by motor E, stabilizes the temperature of the water of the plethysmograph (shown shaded). Pipe G transmits volume changes to the recorder, which may be either a piston as in Fig. 6.16, or a float as in Fig. 6.15; in either case a side tube and syringe as shown in Fig. 6.16 are required for calibration and adjustment. F is a thermometer and the dotted line enclosing the forearm represents the rubber sleeve.

volume which follow immediately after each heart beat. Stop the kymograph. Inflate the pneumatic cuff at the wrist to 200 mm Hg to prevent blood entering the forearm from the hand. This cuff should be kept inflated throughout any experiment but deflated immediately after. Observe the reactive hyperaemia in the hand when the deflation takes place. Wait one minute after the wrist cuff is inflated, then start the kymograph and record a base line. With the kymograph still running inflate the collecting cuff for 20 to 30 seconds by turning the tap connecting it to the air reservoir. The kymograph is then stopped and the tap turned to release the pressure in the collecting cuff. The wrist cuff is deflated and the record inspected. It will have the general appearance shown in Figure 6.17. It shows (a) an immediate 'jump' or artefact due to movement of the tissues at the moment of inflation (the forearm and the apparatus should be arranged so that it is as small as possible); (b) a steadily climbing line which is used to calculate the rate of inflow; (c) a line with a diminishing slope as the rising venous pressure ultimately exceeds the collecting pressure and forces blood back under the cuff towards the heart and in addition opposes arterial inflow.

At the end of the experiments the smoked paper is varnished and the forearm blood flow calculated in the following way. Before removing the plethysmograph make an ink mark around the skin of the forearm at the upper and lower ends of the plethysmograph. Find the volume of the hand including the distal part of the forearm not enclosed in the plethysmograph and then of the hand and forearm by displacement of water in a measuring cylinder. The difference between the two volumes gives the volume of the forearm inside the plethysmograph. On each record of blood flow draw a sloping line through the initial portion of the curve as indicated in Figure 6.17. Find T from the formula $T=60/DV$, where D=vertical distance in cm. described by the writing point of the recorder per 1 ml. increase in fluid content of the plethysmograph (this can be obtained from the calibration scale); and V=volume of the forearm contained in the plethysmograph in 100 ml. Extend the sloping line drawn through the tracing till it cuts the time trace. Measure a horizontal distance of T seconds along the time trace from this point. The vertical distance in cm from the second point to the sloping line above is the blood flow in ml. per 100 ml. forearm per minute.

To calculate the resting forearm flow make a series of ten observations at half-minute intervals. The wrist cuff is left inflated throughout the series but deflated immediately afterwards. The subject rests quietly all the time.

6.49 Blood Flow after Sustained Contraction of the Forearm Muscles

The subject clenches his fists as tightly as possible for one minute. With the kymograph running continuously observations are made at very short intervals (15 seconds) immediately after relaxing the muscles, and then later at somewhat longer intervals until the blood flow returns to the resting level.

6.50 Blood Flow after Rhythmic Contraction of the Muscles

The subject clenches and unclenches his hand vigorously in time with a metronome beating at 80 times per minute for a period of one minute. When the exercise is over observations are made first at short intervals and then at longer intervals until the blood flow returns to the resting level.

6.51 Blood Flow after Occlusion of the Circulation

A second cuff is applied proximal to the collecting cuff and is inflated to 200 mm Hg. for five minutes. Five seconds after this cuff is released a blood-flow observation is made and is followed by observations at intervals of about 15 seconds, until the flow returns to the resting level.

6.52 Blood Flow and Local Temperature

The resting blood flow through the forearm is measured with the water in the plethysmograph at 15° C, 34° C and 43° C. Compare the results.

Possible Sources of Error in Plethysmography

(1). If the collecting pressure applied at the elbow is too low the venous return may not be completely arrested. If too high a collecting pressure is used the arterial inflow is reduced. The pressure which gives the highest inflow should, therefore, be used; 70 mm Hg. is usually the most satisfactory pressure.

(2) The cuffs should be applied as close to the ends of the plethysmograph as possible to get an accurate record. In the case of the forearm, however, the upper end of the plethysmograph is usually placed near the elbow to contain as much of the forearm muscles as possible. The collecting cuff must, therefore, be applied above the elbow and it has to be assumed that the changes in volume of the tissues inside the plethysmograph are representative of the changes which occur between the two cuffs.

(3). The veins should not be filled with blood before the experiment begins. If the veins cannot be easily distended, because of this initial filling, the venous pressure will rise very rapidly and will affect the arterial inflow.

(4). The volume recorder should give a faithful record of the volume changes without any lag.

In general, it may be said that venous occlusion plethysmography probably leads to an underestimate of the true blood flow.

6.53 Hand Plethysmography

A similar series of experiments can be carried out to investigate blood flow in the hand. The light rubber sleeve is replaced by a light rubber glove and only a wrist collecting cuff is needed. Compare the composition of the hand with that of the fore-arm, in respect of bone, muscle and skin. Discuss this in terms of blood flow.

HAND CALORIMETRY

Hand blood flow is almost entirely skin blood flow. Therefore hand blood flow can be estimated by measuring the rate of heat elimination from the hand into a water-filled calorimeter. Accurate results require sophisticated apparatus, numerous pre-cautions and meticulous technique (see 6.55) but the principles involved in the method are well demonstrated in the simpler experiment which is described first.

6.54 Simplified Hand Calorimetry

Apparatus: This consists of a vacuum flask, a thermometer, a stop-clock and a glass cylinder. The flask is large enough to accommodate both the subject's hand and the thermometer without their touching one another. The thermometer is graduated to 0·1° C. The cylinder, also, is wide enough to take the subject's hand. Water close to room temperature is used in the flask instead of water in the ideal range (see 6.55); this minimizes heat exchange with the surroundings and no correction for such exchange is applied in the calculations.

Method: Place 500 ml. of water at about room temperature in the vacuum flask and measure the temperature to the nearest 0·1° C. The subject makes an ink mark round his wrist and immerses his hand in the water in the flask to this level. Start the stop-clock simultaneously. The subject makes slight finger movements to stir the water but must take care not to touch the thermometer bulb and must keep his hand submerged to the mark.

Measure the temperature at intervals of one minute for 12 minutes.

The volume of the hand involved in the heat exchange is determined using the glass cylinder. Fill the cylinder to the brim with water. The subject inserts his hand gently to the ink mark and displaces an equivalent volume of fluid and then withdraws it. The volume of the hand is then that volume of water which must be added to the cylinder to refill it.

During the experiment the hand exchanges heat not only with the 500 ml. of water but also with the flask. The water equivalent of the flask is determined as follows. Add 500 ml. to the flask, leave it for three minutes and then measure the temperature (T_1). Now add a further 300 ml. of water at a temperature (T_2) about 10° C higher than T_1. Stir gently for three minutes and measure the resultant temperature (T_3). Then, if the water equivalent of the flask is E and no heat loss to the environment is assumed,

$$(500 + E)(T_3 - T_1) = 300(T_2 - T_3)$$

Hence E can be calculated and added to the 500 ml. used.

Calculations: Prepare a graph showing Temperature against Time. Find the slope between the 6th and 12th minutes and express it as ° C rise per minute. (Why is the graph steeper at the beginning of the experiment?)

Since 1 ml. of water, or its equivalent, gains 1 calorie when its temperature rises 1° C, the heat gained per minute by the flask and its contents $=(500+E)$ X calories; where E is the water equivalent of the flask and X is the temperature rise per minute in ° C.

Assume that blood enters the hand at 37° C and leaves at the temperature of the water. If the temperature rise of the water has been uniform from the 6th to the 12th minutes, the average temperature of blood leaving the hand during this time is represented by the temperature at the 9th minute. Let this temperature be t° C. Then, since 1 ml. of blood liberates one calorie falling through 1° C, the average ml. of blood flowing through the hand between the 6th and 12 minutes releases (37—t) calories;

and the total heat loss per minute is represented by Q(37—t), where Q is the total hand blood flow per minute.

But, Heat gained=Heat lost.

Therefore, (500+E) X=Q(37—t) and, if V=Volume of hand immersed in ml., the Hand Blood Flow, expressed as ml. per 100 ml. of hand tissue per minute is equal to:

$$\frac{(500+E)\ X}{37-t} \times \frac{100}{V}$$

Substitute your values and calculate flow for your subject.

6.55 Accurate Hand Calorimetry

The temperature of the water used in the calorimeter should be close to the normal skin temperature to avoid vasomotor changes. Working the calorimeter at this temperature makes it important to have a very good thermal insulation otherwise heat losses will be significant compared with heat gains from the circulating blood. A temperature between 29 and 32° C is recommended. The rate of heat elimination from the hand into a calorimeter filled with water in this temperature range is proportional to the blood flow through the hand. It is therefore, a convenient, though indirect, method of making observations on blood flow and the factors which control it. A team of at least four persons is required. The experiment should be planned and definite duties allocated to each member of the group.

The calorimeter is constructed from a 1-gallon vacuum flask (Fig. 6.19). A square of thick plywood with an underlying layer of sorbo rubber is clamped over the open end. The hand passes through a hole cut in the plywood and through an oval aperture cut in the sorbo rubber which fits loosely around the wrist. The calorimeter has a water equivalent of 200 ml. and contains 3500 ml. of water. A paddle, driven by a small electric motor attached to the top of the plywood, rotates in a copper tube and stirs the water. Water is drawn into the tube from the lower part of the calorimeter and is discharged by centrifugal force through a number of holes drilled in the tube below water level.

The temperature of the water is measured with a thermometer graduated in hundredths of a degree centigrade. This passes through a hole in the plywood and rubber, and is held at the top by a clamp attached to a vertical steel rod bolted to the plywood. Each thermometer reading should be taken with the observer's eye at the level of the top of the mercury column. Readings will be accurate to 0·01°C and can be judged to the nearest 0·001°C.

If the temperature rises T thousandths of 1°C (or T milli-degrees C) per minute with the hand in place and if, when the hand is removed, it falls t millidegrees C per minute due to loss of heat to the surroundings, the rate of heat elimination from the hand is $\frac{(T+t)}{1000} \times 3700$ cal per min.

The cooling correction of the calorimeter is first obtained. The calorimeter will be found filled with water between 29 and 30° C. Start the stirring motor, close the hole for the hand at the top with the wooden block and carefully insert the thermometer. Take readings every minute for 10 minutes and calculate the average temperature fall per minute (t millidegrees C).

During this period the subject removes his jacket and waistcoat and with his feet and arms bare is cooled by standing in front of an electric fan. The mouth temperature should be taken now and every five minutes throughout the experiment with a clinical thermometer. Immediately after each observation the mercury is shaken down and the thermometer is replaced under the subject's tongue. He then sits on a chair and inserts one hand into the calorimeter. It is important to get the subject into a comfortable position and to put the calorimeter at a suitable height so that the arm can be relaxed. The wrist should be at the level of the hole in the lid and an ink mark is made on the wrist so that the constancy of the depth of immersion of the hand can be checked every few minutes. Cotton-wool is packed loosely round the wrist to prevent air currents. The immersed fingers are held slightly apart to allow the water to circulate. Thermometer readings are made every minute, and the rise in temperature each minute, T millidegrees C, calculated. For the first 10 to 15 minutes the thermal exchange between the hand and the water depends mainly on the initial temperature of the hand. If the hand is hot at the beginning of the experiment the calorimeter warms up but if the hand is cold the calorimeter cools down. Until the hand has come into thermal equilibrium with the calorimeter, temperature changes due to this cause may greatly exceed those due to heat released from the warm blood circulating through the hand.

When the temperature rise each minute has been reasonably steady for 10 minutes (usually 20 to 25 minutes from the start of the observations) the subject places his feet in a water bath at 42 to 43° C and a blanket is draped round him. The water bath is maintained at this temperature by a bunsen burner and must be continuously stirred with a wooden spoon. This temperature is the hottest that can be used without discomfort. After an interval, usually about 10 minutes, the skin vessels dilate (due to a reduction of vasoconstrictor tone) and a greater quantity of blood flows through the hand. The dilatation is signalled by an increase in the rate of temperature rise of the calorimeter.

After a large rate of temperature rise (about 70 to 100 millidegrees per minute) has been maintained for about 10 minutes, the subject takes his feet out of the water bath and the blankets are removed. Observations are continued until the rate of the temperature rise in the calorimeter is reduced to about its previous low level.

Make a graph by plotting the temperature rise T, in millidegrees C per minute on the ordinate (1 cm = 10 millidegrees C) against time in minutes on the abscissa (2 mm = 1 min). Draw a base line at a distance below the zero equal to the cooling rate t, in millidegrees C per minute. The height of the curve above the base line shows the effect of the immersed hand on the calorimeter temperature (Fig. 6.20). The mouth

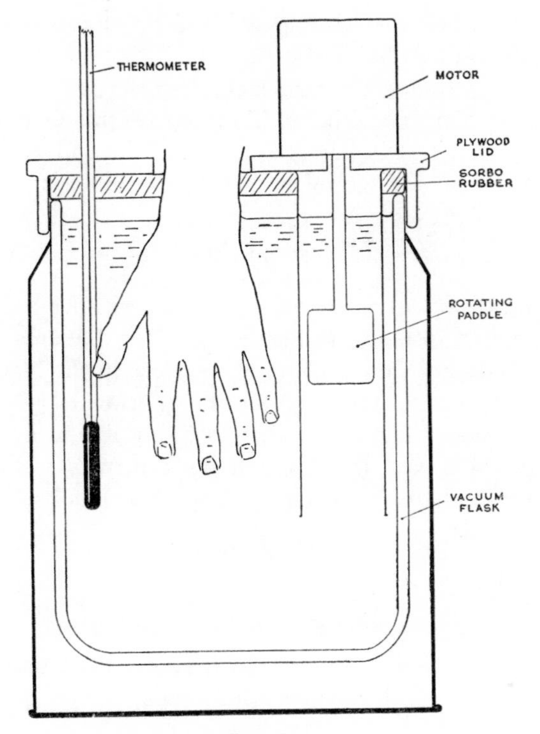

FIG. 6.19

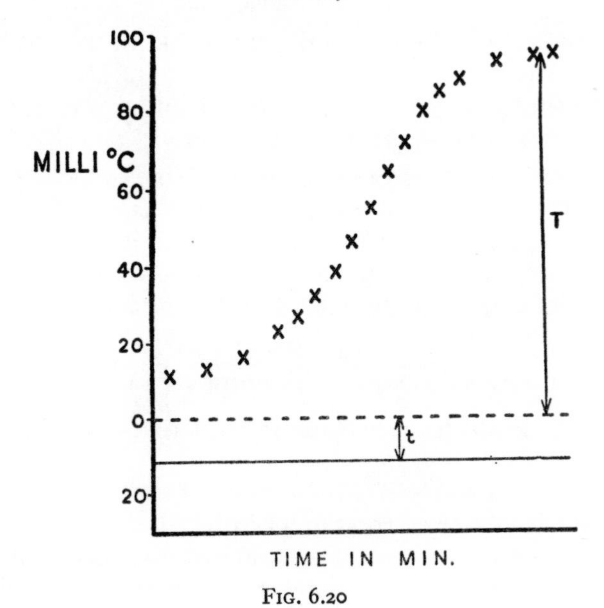

FIG. 6.20

temperature may be shown on the same graph, and the duration of immersion of the feet in the hot water should also be indicated.

The results are more frequently expressed, however, as the number of calories eliminated by 100 ml. of hand per minute. To calculate this determine the volume of the immersed hand, as described in paragraph 6.54, at the end of the experiment. The rate of heat elimination in cal per 100 ml. of hand per minute is:

$$\frac{T+t}{1000} \times 3700 \times \frac{100}{\text{Hand volume in ml.}}$$

A subject in whom the sympathetic nerves to the arm have been divided shows no vasodilation in the hand in response to heating the feet. Patients with disease of the peripheral blood vessels do not show as large an increase in the blood flow through the hand, and hence the rate of temperature rise in the calorimeter is less. The mechanism of release of sympathetic tone is not fully understood, but it is likely that the stimulation of heat-sensitive nerve endings in the skin and the return of warmed blood from the feet to the temperature-regulating centre in the hypothalamus both play a part.

Failure to get a normal response may be due to:

1. Inadequate heating.
2. Failure of the mechanism which appreciates the heating, peripheral or central.
3. Failure of, or absence of, the sympathetic nerves to the blood vessels in the hand.
4. Inability of diseased vessels in the hand to dilate.

Where greater accuracy is required allowance must be made for two variables:

1. The water equivalent of each calorimeter should be individually determined. This is done by adding to a known mass of water at a given temperature already in the calorimeter a known mass of water at a different temperature and determining the temperature of the resulting mixture.
2. The amount of heat released from a given volume of blood (and therefore the value of T+t) is proportional to the difference between blood temperature and calorimeter temperature. If the calorimeter is used within a narrow range of temperature, neglect of this variable does not lead to a large error. The heat released may be adjusted by multiplying by the factor:

$$\frac{37-32}{37-\text{Calorimeter Temperature } ^\circ\text{C}}$$

This gives the heat release which would occur if the calorimeter temperature were always 32° C.

6.56 The Effects of Circulatory Arrest to a Limb

Both the subject and the observer must read over the detailed description of this experiment beforehand to obtain a clear understanding of the procedure. The effects

of circulatory arrest include colour and temperature changes and transient derangements of sensory and motor nerves. Make a note of the times of onset of physical signs and sensations and give a brief description of them. If possible chart the loss of touch, pain (pin-prick), position-sense and motor loss. This experiment affords both the observer and the subject a practical experience of certain peripheral disorders.

The arm should be warm at the beginning of the experiment. It is supported on the bench and kept quite still and relaxed, otherwise muscular pain will develop before the experiment is completed. A sphygmomanometer cuff is placed around the upper arm, inflated to 200 mm Hg. and maintained at this pressure by pumping as necessary. The events to be noted can usually be described as occurring in four periods.

First Five Minutes. The limb temperature begins to fall immediately and at the end of this period the finger temperature is noticeably lower in the occluded hand. The observer should test this with the back of his fingers. Compare the two hands of the subject. The skin gradually becomes blue in colour and may be violet by the end of the first period. A few islets of red which may persist are due to minute trickles of blood passing through the nutrient arteries of the bones to supply small areas of skin.

Second Five Minutes. The skin temperature continues to fall and white patches may appear in certain regions. The cause of this local vasoconstriction is unlikely to be oxygen lack since this usually causes capillary dilation.

Third Period. After 13 to 15 minutes, disturbances of the function of peripheral nerves are detectable. Touch sense is the first affected and is lost from the tips of the fingers upwards; this can be demonstrated with a von Frey hair (7.2). At this stage pain and temperature sense are still present but they gradually decline from the extremity upwards. Test for loss of position-sense by moving the subject's finger and asking him to tell if it is bent or straight. The motor nerves supplying the thenar muscles begin to fail about the 20th minute, and voluntary movement of these muscles fails about the time that touch anaesthesia has reached the wrist. In testing for motor paralysis the muscles must be moved as little as possible, otherwise ischaemic pain will terminate the experiment. As a rule the extensors of the wrist and fingers become paralysed about the 30th minute and if the forearm is raised from the bench a typical 'wrist-drop' can be demonstrated. Stop the experiment at this point or at the 30th minute whichever comes first.

Recovery Period. On release of the cuff recovery is very rapid. The whole limb flushes bright red (reactive hyperaemia) and nerve recovery is almost complete in one minute, finger-tip touch recovering last. Acute 'pins and needles' sensation is usually felt in the distal part of the limb for several minutes but it passes off as the vascular and nerve conditions stabilize.

In many cases the experiment has to be terminated prematurely because of ischaemic muscular pain (10·7). This may be due to inadequate relaxation of the arm or too frequent testing of the muscles. An arm must not be subjected to circulatory arrest more than once in the course of one day.

CHAPTER SEVEN

SENSORY PHYSIOLOGY

CUTANEOUS SENSES

There are several types of nerve ending in the skin, and they respond to stimulation with a discharge of nerve impulses in the sensory nerve axons connecting them by way of the dorsal roots to the spinal cord. The first experiment in this section is concerned with the direct observation of nerve impulses originating in the mechanoreceptors of the skin of the frog, the remainder of this section is concerned with human cutaneous sensation, which is an extremely complex matter, since it involves not only the activity of the sensory nerve endings but processes in the central nervous system. Very little is known about the nature of these processes.

7.1 Sensory Nerve Impulses from the Frog's Skin

Stun and pith a frog. Make a mid-line incision through the skin overlying the dorsal lymph sac (Fig. 7.1). Note that very slender nerves traverse the lymph sac, connecting the vertebral column with the inner surface of the skin. Cut out a piece of skin, about 1 cm square, together with its supply nerve. Detach the nerve close to the vertebral column and tie a cotton thread round the cut end. Tie out the piece of skin, inner surface uppermost, to a glass rod frame and immerse in a Petri dish containing Ringer's solution (Fig. 7.2). Adjust a silver foil ring over the nerve and fill the ring with paraffin oil so that the nerve can be lifted up into the oil and placed on a silver wire electrode. The electrode and ring are connected to the input terminals of a high gain differential amplifier. The output of the amplifier is connected to a cathode-ray oscilloscope and sound monitor unit with loudspeaker. Explore the skin mechanically with a slender glass rod. Note that when certain regions are stimulated action potentials occur in the nerve. See if you can distinguish between sensory endings in terms of their adaptation rate to a constant stimulus or in terms of the action potential size on the oscilloscope screen. At the end of the experiment show how a local anaesthetic depresses and then extinguishes the activity of sensory nerve endings by adding a solution of procaine hydrochloride to the bath so as to produce a final concentration of 0·2 per cent.

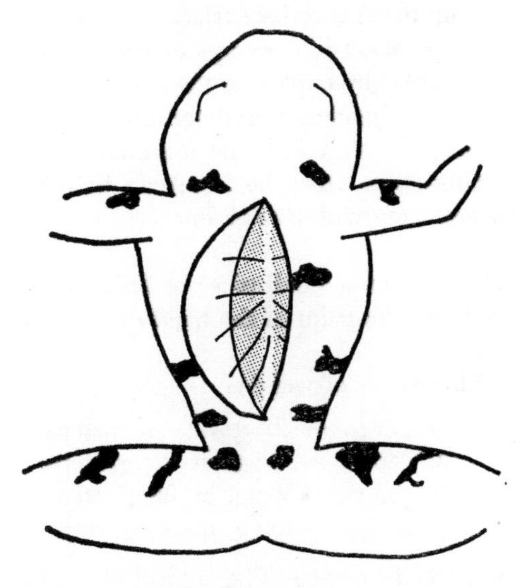

Fig. 7.1
Dissection of dorsal lymph sac.

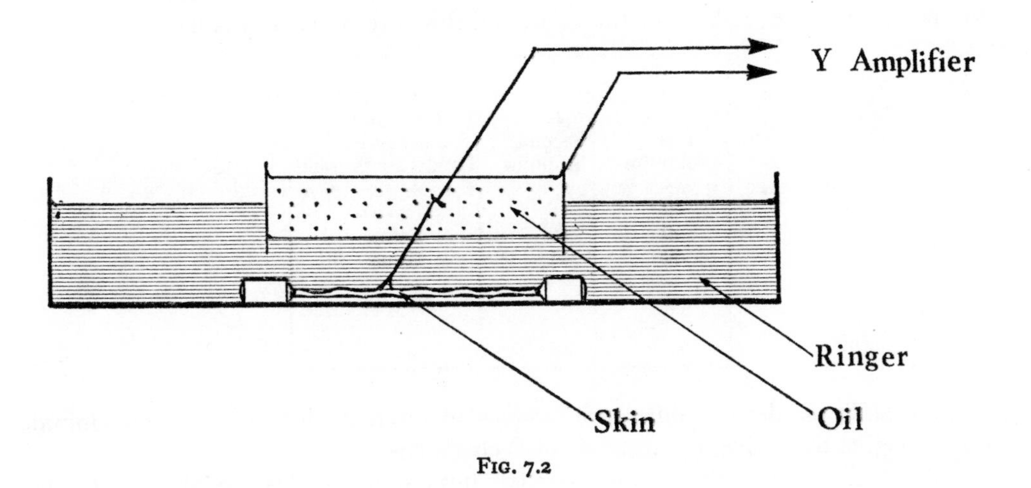

Fig. 7.2

7.2 Human Cutaneous Sensation

Apply the rubber stamp to (a) anterior surface of distal phalanx of thumb, (b) back of hand, (c) back of forearm; also make stamps in your book for records. Make a map of the touch spots in these areas by exploring them with von Frey's bristles—mounted hairs of known thickness which bend at a predetermined pressure. The observer pushes the bristles against the subject's skin with the minimum pressure which will produce bending and the subject indicates when he is certain he feels a contact. Explore first of all with a fine bristle and then with a thick one.

7.3 Compare the sensations when a small piece of loose cotton-wool is lightly stroked over the back of the hand and the palm of the hand (a hairy and a hairless part).

7.4 You are provided with a series of weights

1. 1 g. — 20 g. in 1 g. steps
2. 20 g. — 80 g. in 5 g. steps
3. 80 g. — 200 g. in 20 g. steps
4. 200 g. — 800 g. in 100 g. steps

The actual mass is marked on each. The subject closes his eyes and lays his hand on the bench *palmar surface up* with the fingers a little apart. The dorsal aspect of the fingers must be kept in contact with the bench throughout the experiment. The observer selects a pair of weights and places them successively on the middle phalanx of the subject's middle-finger, each for three seconds. The subject reports whether the second stimulus was heavier, lighter, or the same, and this is recorded as a table:

1. First Stimulus	2. Second Stimulus	Answer: 2 is lighter 2 is heavier 1 and 2 are the same
40	48	Heavier
40	40	Same
40	41	Same
—	—	—
38	42	Heavier
—	—	—
—	—	—

The ability to detect a difference between two weights depends on the magnitude of the weights themselves (an example of Weber's law).

Test each general weight range separately but use the weights irregularly and sometimes use the same weight twice. Make a large number of observations and determine how the minimum detectable increment varies with weight range for increments of 1 g., 5 g., 20 g., and 100 g., i.e. find the greatest weight at which you can consistently detect an increment of 1 g., and so on.

7.5 With the aesthesiometer (dividers) test the subject's discrimination. Apply both points **simultaneously** and lightly. Test the back of the hand, the palmar surface of distal phalanx of the thumb and the back of the forearm. Vary the distance between the points irregularly. Record the distance in one column and the subject's answer—'one', 'two', 'don't know'—in another column. Determine the smallest interval appreciated with certainty on the three areas. Measure the distance apart of the touch spots in the diagrams obtained previously. What relation do these distances bear to the figures obtained now?

7.6. Aristotle's experiment. The subject closes his eyes and crosses the middle finger over the index finger. The observer lays a pencil in the space between the finger-tips. The subject describes his sensation.

7.7 With your fine forceps investigate the qualities of sensation that can be aroused from a single hair on the back of your hand. Try to decide if the endings are rapidly or slowly adapting. Does a steady deflexion of a hair produce a continuous sensation? Close your eyes and ask your partner to move a single hair in various ways without touching your skin and see which movements you can detect. Consider the value of long projecting hairs (vibrissae) on the head of a nocturnal animal such as a cat or rat.

7.8 Cutaneous Localization

The subject shuts his eyes. The examiner lightly touches the skin of the subject at various points. The subject then puts his finger on the part touched. By an extension of this simple localization the direction of lines drawn on the skin can be recognized. The subject closes his eyes and holds out his hand palm upwards. The examiner traces out on the palm with his finger, or a blunt pencil, letters or figures. The subject attempts to identify them. Information can be readily transmitted via cutaneous endings in densely innervated zones, such as the finger-tips, and this is the basis of a reading method for blind persons.

7.9 Sense of Pain

Using a needle make a map of the pain spots in the same areas as explored in testing the sense of contact.

From these and previous experiments could you decide what would be a suitable area to administer a hypodermic injection?

7.10 Varieties of Pain

Prick yourself with a needle. Pull on a hair for a short time, then for a longer time. Describe the resulting sensations. Compare with the sensation produced when (1) the web of skin between the 3rd and 4th fingers is compressed, or (2) the tendo calcaneus is compressed.

7.11 Hyperalgesia from Injury

Hyperalgesia means increased sensitivity to pain. When the skin is injured by any means, mechanical, heat, cold or ultra-violet light, a state of hyperalgesia develops in the injured skin and the area immediately adjacent. The hyperalgesia does not appear at once but develops about an hour after injury and persists for many hours or days. Familiar examples are sunburn and chilblains. The underlying cause of the hyperalgesia is unknown; it has nothing to do with the 'triple response' which is also a sequel of injury.

With a needle scratch 1 sq. cm of skin on the forearm making 10 vertical and 10 horizontal scratches, using a pressure just short of drawing blood. Alternatively, pick up a tiny fold of skin with artery forceps and crush it.

When the initial pain has died away, note that the area is not unduly sensitive to pain—test by drawing a needle over it. Note the 'triple response' to injury—red reaction, flare and weal (p. 170). Re-examine 1 to 2 hours later and test for hyperalgesia in the following ways—(1) draw a needle point along the forearm crossing the injured area en route, and note that pain is produced in the injured area with a pressure that is insufficient to arouse it from the normal skin; (2) apply a test tube containing water at 45-50° C—note the violent stinging character of the pain; (3) stroke firmly with the finger.

Make observations at intervals during the evening and on the following morning and record the duration of (1) the redness; (2) the swelling and (3) the hyperalgesia.

Temperature Sense

7.12 Prepare three jars of water in the order hot, warm, cold. The water in the first should be as hot as can be borne by the finger without pain. Dip the right forefinger in cold, the left in hot. After one minute place both together in warm. Record your sensations. The human temperature sense is relative and not absolute; in this experiment it appears to depend on the abstraction of heat from, or of addition of heat to, the skin.

7.13 Compare the temperature sensations given by a piece of flannel and a piece of metal at room temperature. Try to explain this on the basis of the physical changes occurring in the skin.

7.14 Use the rubber stamp to map out an area on the back of the hand between any two metacarpals. Apply the stamp to your book. Trail the chilled blunt-ended brass rod gently across the area mapped out. Mark on the diagram in your book spots at which a cold sensation is appreciated. Heat the brass rod to about 45° C and again explore and record the 'hot' spots. Repeat these observations after the hand has been held in (a) hot water and (b) cold water, for three minutes. Are the 'spots' fixed in position?

Do the pain, temperature and touch spots coincide?

7.15 Use two large glass test-tubes containing water at 36° and 38° C. Apply these to the face of the subject in turn. He will be able to tell easily which is warmer. What is the minimum perceptible difference?

7.16 Vibration Sensibility

Vibration sensibility is usually tested by applying the base of a vibrating tuning-fork of low pitch to a subcutaneous bony prominence. The subject indicates how long the sensation lasts.

To obtain standard results a special tuning-fork has been devised. A small notched plate is fixed to the inner aspect of the end of one prong, the notches are overlapped on both sides by a double plate fixed to the inner side of the other prong so that the notches are normally hidden. Owing to the separation of the prongs when the fork is vibrating strongly the notches will be seen as a 'window' which gradually diminishes in size as the amplitude of vibration decreases. The time in seconds between the disappearance of the 'window' and the end of the sensation of vibration appreciated by the patient is taken as a measure of the vibration sensibility.

Record the times found in your case for the radial styloid and the external malleolus and compare with the times found by other members of the class.

ELECTROPHYSIOLOGY OF THE FROG MUSCLE SPINDLE

7.17 Response of the Frog Muscle Spindle to Stretch

Detach the foot from a pithed frog by sectioning the bone mid-way between knee and ankle. Remove the skin carefully and identify the peroneal nerve. Pin out the foot in the electrically screened bath and place the peroneal nerve on the recording electrodes. Cover the foot and nerve with paraffin oil to prevent drying. The recording electrodes are connected via a pre-amplifier to a loudspeaker (or headphones if the laboratory is noisy) and to a cathode-ray oscilloscope. The overall maximum sensitivity required for the oscilloscope is 100 μV per cm. Probe the dorsal muscles of the foot with a glass rod and note the modulation of the discharge of sensory nerve impulses. Select a small muscle, e.g. the interosseous (M.ext.brev. dig. 3 or 4) (see Fig. 7.3), detach the distal tendon and attach a thread. Pull on the thread and so stretch the muscle. Note that the initial effect declines, i.e. the endings show adaptation. To obtain a simpler record, dissect the peroneal nerve closer to the muscle under examination, cutting the nerve filaments to adjacent muscles. With care a preparation containing only a few active fibres can be obtained.

7.18 Dissection of the Extensor longus dig. 4 Muscle of the Frog

This slender muscle is about 15 mm long and consists of about 60 extrafusal muscle fibres running parallel from tendon to tendon. Amongst the extrafusal fibres are two or three bundles of intrafusal fibres with stretch receptors mounted on them. Removal of the muscle together with a length of nerve calls for some skill in dissection, but with practice it can be achieved in less than an hour. A binocular dissecting microscope, good lighting and a steady hand are required.

Cut off the leg from a pithed frog and remove the skin from the knee distally. Pin the leg on a cork board under the microscope. Detach the peroneal and anterior tibial muscles from the knee and dissect them away as far as the ankle tendons. The peroneal nerve will

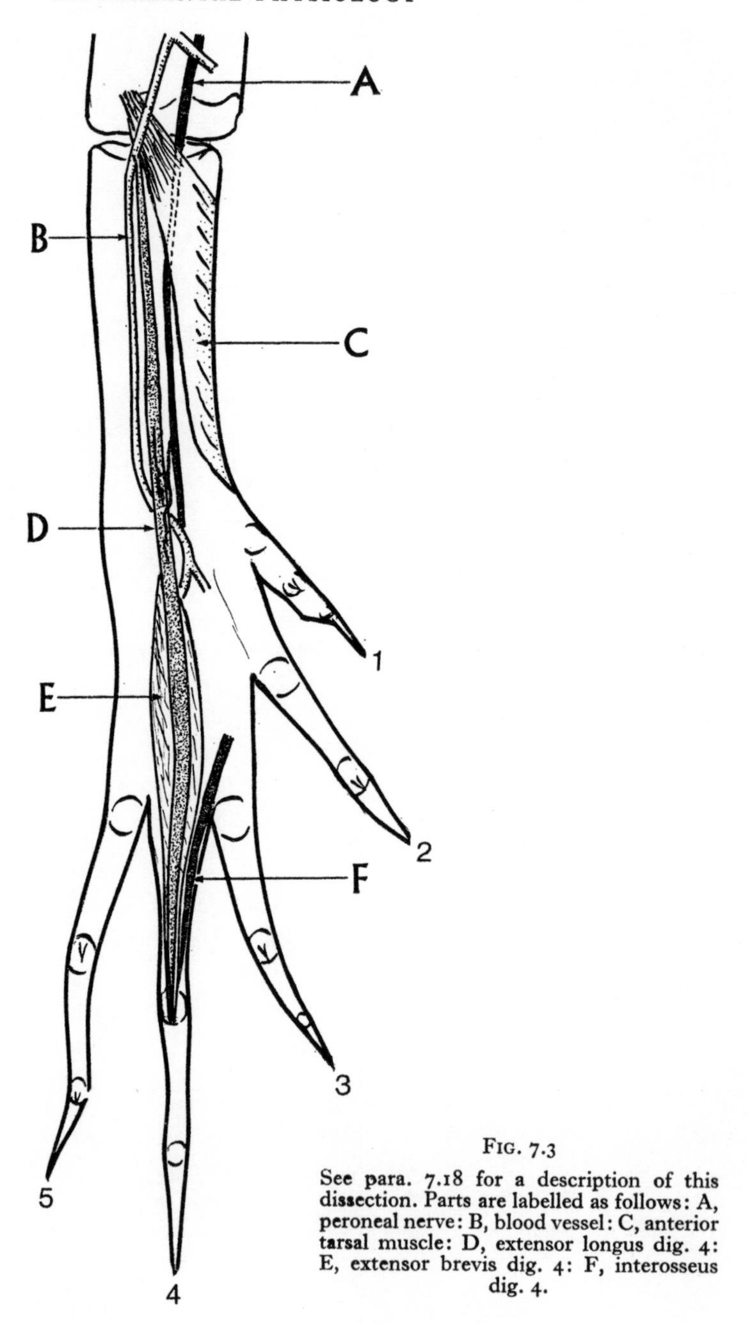

FIG. 7.3

See para. 7.18 for a description of this dissection. Parts are labelled as follows: A, peroneal nerve: B, blood vessel: C, anterior tarsal muscle: D, extensor longus dig. 4: E, extensor brevis dig. 4: F, interosseus dig. 4.

now be visible as it passes under the lateral tendon of the gastrocnemius muscle at the knee and then courses beneath the anterior tibial muscles to the ankle. Sever the nerve at the knee and free it as far as the ankle. Now examine the dorsum of the foot. Beneath the skin, there is a semitransparent layer of connective tissue which lies on top of the toe muscles. See diagram 7.3 and identify the position of the superficial blood vessel, which is usually pigmented. Carefully slit the connective tissue layer over the anterior tarsal muscle and reflect it to one side. You can now see the long extensor muscle of the 4th toe (the 4th is the longest toe) running parallel and close to the blood vessel and extending from a tendon shared with the anterior tarsal muscle proximally to a shared medial tendon at the joint of the 4th toe. The nerve supply approaches the muscle from the lateral peroneal nerve which lies below it, and enters the muscle in the proximal half. Cut across the anterior tarsal muscle and allow the stumps to retract. You can now see the lateral peroneal nerve lying under the extensor longus dig. 4 with a very delicate nerve connecting them. This nerve crosses the blood vessel and is in sharp contrast as it does so. Detach the distal tendon of the muscle from the toe together with a fragment of underlying toe muscle (ext. brevis medius dig. 4) (this fragment makes it easier to secure the end of the muscle since the tendon is very slender) and free the muscle up to the point where you have seen the nerve supply. This nerve contains about 12 fibres of which three or four are sensory axons each connected with one or more spindles. Sometimes it is in two parts. Cut the lateral peroneal nerve distally to the point where the branch leaves it. Sever the proximal tendon of the muscle and transfer the muscle together with peroneal nerve to a dish filled with Ringer's solution.

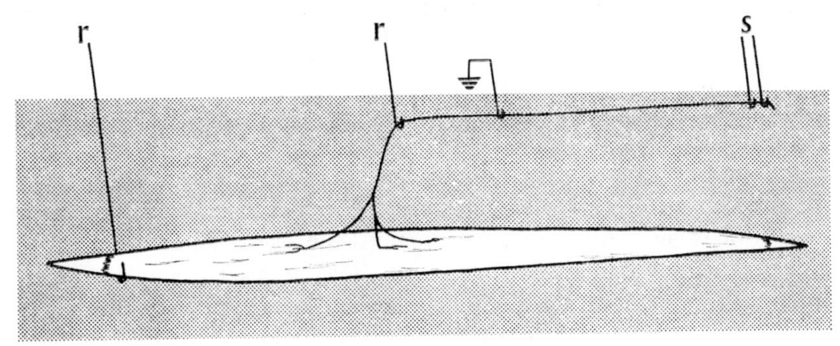

Fig. 7.4
The extensor longus dig. 4 is set up in oil in the bath shown in Fig 2.8 and secured under slight tension applied to the tendons. The electrode arrangement is shown above; (rr) recording electrodes, (s) stimulating electrodes.

7.19 Stimulation of Fusimotor Fibres

Set up the ext. longus dig. 4 muscle and nerve in the small screened plastic bath under the binocular dissecting microscope. The muscle and nerve are immersed in paraffin oil. Arrange the stimulating and recording electrodes as shown in diagram 7.4. The time-base of the oscilloscope is triggered by the stimulator (2CROS), nerve and muscle action potentials are picked up by wire electrodes and after amplification displayed on the Y axis of the oscilloscope. A loudspeaker unit enables the impulse discharge to be heard. The total amplification required is 100 μV per cm Y axis deflection. Adjust the time-base velocity to 20 msec/cm. First confirm that the frequencies of discharge of sensory nerve impulses is influenced by muscle stretch. Then stimulate the nerve at a frequency of one shock per second, pulse length 0·5 msec. Adjust the stimulus strength until each shock causes a muscle twitch. The muscle is prevented from shortening by the clamps attached to the tendons, i.e. the contraction is isometric. Note that each shock is followed by a burst of sensory impulses from the spindle. These impulses are provoked by contraction of the intrafusal muscle in the spindle.

The neuromuscular junction of intrafusal muscle is less sensitive to the action of curare than the neuromuscular junction of extrafusal muscle. It is possible, by adjusting the dosage, to block extrafusal contraction while preserving intrafusal contraction. In these circumstances no general twitch is seen but with a good microscope the spindles can be seen to contract and this corresponds in time to the burst of sensory impulses in the nerve.

With a syringe apply a few drops of 8×10^{-7} tubocurarine in Ringer to the muscle. Stimulate occasionally during the next ten minutes to follow the development of the neuromuscular block. If extrafusal block is not complete, increase the curare concentration, e.g. to 10^{-6} or $1{\cdot}5 \times 10^{-6}$. Curare can be readily washed out of the preparation with Ringer's solution.

The record from the preparation will change as the curare takes effect, the large deflection due to the extrafusal electromyogram declines and disappears leaving the burst of sensory impulses from the spindles provoked by the contraction of the intrafusal muscle.

PROPRIOCEPTIVE SENSES IN MAN

Although proprioceptive sensation normally plays a very large part in maintenance of posture and control of movement, extensive defects can be well-concealed by compensatory use of other sensory systems, mainly the eyes. Therefore tests performed with the eyes shut may unmask latent defects. The simplest example of such a test is the ability to stand erect with the heels together and eyes closed; a normal subject may sway slightly, but a subject with proprioceptive loss frequently sways violently and may even fall. Not all tests, however, require that the eyes be closed, e.g. assessment of the ability to walk a straight line.

7.20 With eyes shut the student describes the shape of various objects placed in his hand. He is asked to name familiar objects such as a pen-knife, a rubber, a piece of string placed in his hand. Ability to recognize objects in this way is called stereognosis. He attempts to copy with one hand the position of the other hand and fingers set by the observer. He stretches out his right arm to the right and then quickly bends the arm and attempts to touch the point of his nose with the tip of the right index finger. This is repeated for the left arm.

Test the subject's ability to walk along a straight line on the floor. Now stand the subject erect, with heels together and eyes shut (Romberg's test). Note the slight, normal sway. Why does this occur? Standing on one leg is a difficult procedure, and an unfair test in the elderly, but the normal young subject should accomplish it, although sway becomes more noticeable.

The subject, with eyes open, sits opposite the observer, stretches out his right arm directly in front and touches the observer's finger with his index finger. He now closes his eyes and moves his arm back till it is pointing to the right; he then attempts to return the arm to its original position with the index finger touching the observer's finger. Repeat with the other arm. Repeat these manoeuvres in a vertical plane. If the subject cannot touch the observer's finger but comes to rest some distance away this is described as *past pointing*.

7.21 The subject studies the dynamometer, and makes a few trials noting the excursion on the gauge. Then he closes his eyes, compresses the spring, and gives his estimate of the force of compression. With the eyes open note the muscular tension on bringing the pointer to any convenient number. Try to bring it to the same number with the eyes closed.

7.22 The subject spins round rapidly for half a minute. When he stops he looks at the observer's nose. The observer studies his eyes for movements and describes their character and direction. These movements are called post-rotatory nystagmus. During the time that they are present disturbances of locomotor skill occur because false information is being fed into the nervous system by the labyrinthine mechanism. This false information gives the subject a sensation of rotation even though he is standing still. Test the subject's skill by asking him to touch a chalk mark on the wall at shoulder height, first with his eyes open and then with them closed. Does past pointing occur? Test his general locomotor skill by asking him to walk along a chalk line on the floor, but stand by him to prevent any accident. Note the time taken for recovery of normal skill. The subject should describe his sensations and test the loss of his own visual skill by attempting to read the clock on the laboratory wall during the rotatory and post-rotatory phases.

7.23 The subject sits on a chair mounted on a turntable and is spun steadily for half a minute. His head should be bent forward by 30° towards his chest so as to give maximum stimulation to the horizontal semicircular canals. Rotatory nystagmus occurs at the beginning of the rotation and at a rate to match the spin, but this response declines and reappears in a reversed form as post-rotatory nystagmus when the chair is stopped. Test the subject's skill and record his eye movements as in paragraph 7.22

7.24 Caloric Test of Labyrinthine Function

(Do **not** apply this test if the ear drum is perforated).

This test must be carried out under the supervision of a member of staff.

Only the lateral semicircular canal is near enough to the external auditory meatus to be affected by changes of temperature. The subject lies on his back with a pillow below his head to tilt it forward about 30°. In this position the canal is vertical and the reactions to stimulus are most obvious.

Fit a filter funnel to a small catheter or a narrow rubber tube and run water at either 30° C or 44° C (i.e. either 37—7° or 37+7°) into the meatus for 40 seconds. Catch the water flowing away in a kidney dish placed against the head. Cold water induces a flow of endolymph from the ampulla which is placed anteriorly. This is also the direction of flow when the subject suddenly stops rotating to the right. As in the rotation experiment observe the character of the nystagmus and its duration; examine any tendency to past point or to fall on standing erect. Irrigation with water at 44° C has exactly the opposite effect.

CHEMICAL SENSES

7.25 Taste

You are provided with (a) 5 per cent. cane sugar—sweet; (b) saturated quinine sulphate—bitter; (c) 0·5 per cent. H_2SO_4—sour; (d) 1 per cent. NaCl—salt. The subject protrudes his tongue and dries it. The observer, using small pieces of blotting-paper, paints the various solutions in turn on the tip, side, centre and back of the tongue. Record the subject's sensations on a diagram, noting also the time between the applications of the stimulus and his sensation. Tap the tip of the tongue with the point of a pencil—describe your sensation. Using non-polarizable electrodes (Ag, AgCl) and a 6-volt battery stimulate different areas of the tongue and describe the sensations aroused.

7.26 Smell

The subject closes his eyes, pinches his nose and opens his mouth; the observer places pieces of potato on the subject's tongue. These are replaced in turn by pieces of onion. The substances may be rolled over the tongue but should not be chewed because the texture is different in the two cases. The identification is noted in several tests and control experiments should be made with the nose open. Compare the results of this experiment with the so-called loss or impairment of taste during a 'cold in the head'.

A series of bottles containing well-known odoriferous substances like camphor, cloves, peppermint is available. The subject, with eyes closed, sniffs at each and attempts to name the odour.

HEARING

7.27 The Frequency Range of the Human Ear

A group experiment for 10 to 20 subjects.

The output from a sinewave oscillator is fed into a loudspeaker. The frequency of the oscillator is swept from 10 to 20,000 Hz at a comfortable loudness. Each member of the group of subjects signals with his hand when he can hear the sound. The frequency range of each subject is noted. Indicate on a single diagram the performance of your group.

Generally speaking, the frequency range of hearing becomes more restricted with advancing age (presbycusis); is there any support for this in your own group? As a check on the performance of the loudspeaker it is useful to set up a microphone connected to the Y amplifier of a cathode ray oscilloscope amonst the subjects so that the sound output of the speaker can be seen even when it is inaudible to some subjects.

Sensitivity of the ear

7.28 The subject blocks one ear with a finger, the observer brings up a watch behind his head and notes the distance at which it is first heard; this is then repeated with the other ear. Compare your results, using the same watch, with other subjects.

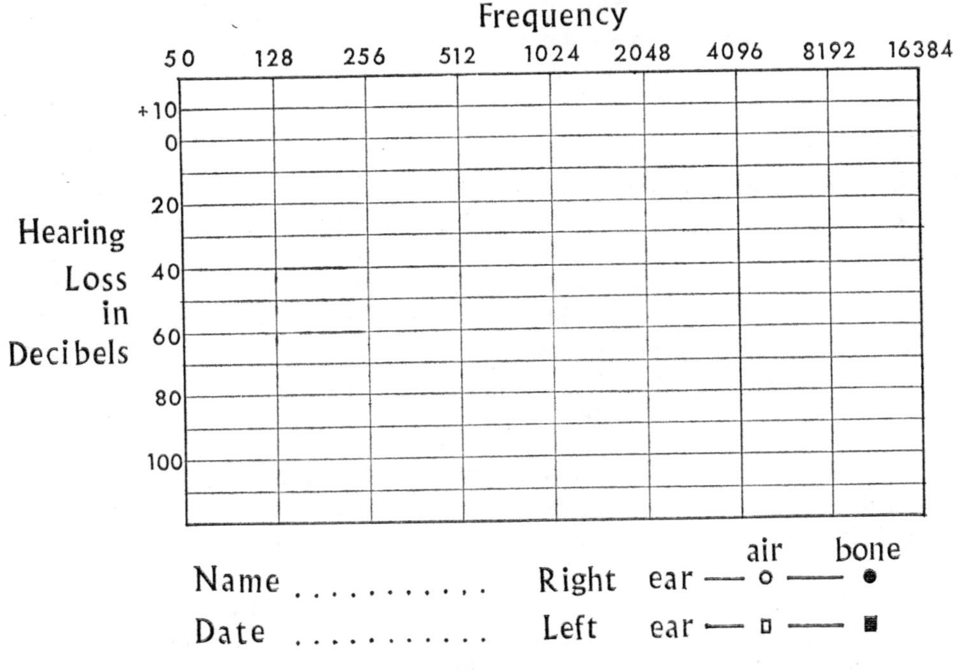

FIG. 7.5
Use this chart to record the results of audiometer test described in para. 7·29.

7.29 Pure Tone Audiometer

This apparatus is used to determine the threshold sensitivity of the ear to pure tones in the range 50 to 16,000 Hz. It is calibrated in terms of frequency and sound intensity and provision is made for the sound vibrations to reach the ear either through the external auditory meatus (air conduction) or through the bones of the skull (bone conduction).

Examine the instrument and note that one dial controls the frequency of the tone and the other the sound intensity. The graduations on the intensity dial indicate the intensity in relation to the threshold intensity of an individual with normal hearing. The normal threshold intensity varies with the tone employed and provision is made for this in the graduations on the dial.

Switch on the instrument, set the frequency dial at 256 Hz and put on the headphones. Adjust the intensity knob until the sound is heard in one ear at a comfortable level. Note the switch on the panel which routes the sound either to the left or right ear. Now reduce the loudness until the sound is just inaudible, note the dial reading and then, moving the dial the other way, increase the intensity until the threshold is reached once more. After a few trials mark the threshold level on Figure 7.5. Repeat the procedure at other frequencies.

If your hearing is normal the graph will be parallel and close to the zero line.

To measure bone conduction, take off and unplug the headphones. Fit the vibrator so that it presses against the mastoid process. Plug the vibrator into the instrument. Note that a different set of graduations is used on the intensity dial and the frequency range is restricted. Measure your hearing thresholds throughout the range available and plot on Figure 7.5.

Conductive and Perceptive Hearing Loss

Sound waves are normally transferred from the outside world to the cochlea by way of the external auditory meatus and the middle ear bones. If a subject's sensitivity to sound is low, this loss may be due to a defect in the external ear, such as wax blocking the meatus, or to some defect of the middle ear and this is called conductive deafness. Alternatively there may be no conductive defect but instead a loss of sensitivity in the cochlea or a more central defect; this is called perceptive or nerve deafness. It is easy to distinguish between sensitivity to air conduction and bone conduction (which by-passes the middle ear) with a pure tone audiometer as described in section 7.29 and since with a calibrated instrument the sound thresholds may be read directly from the dial in terms of the normal threshold, it is possible to decide the amount of conductive or perceptive loss present There are however two useful tests demanding only a tuning fork which were developed before the availability of the pure tone audiometer. The first, Rinne's Test (7.30) compares hearing by air and bone conduction so as to distinguish between conductive and perceptive deafness. The second, Weber's Test, (7.31) uses the fact that an ear with normal perceptive threshold but with a conductive defect shows an improved sensitivity to bone-conducted sounds. Thus if sound is conducted to both ears by bone vibrations from some equidistant point such as the forehead or vertex, provided that both cochlea are normal, it will be heard more loudly in an ear with a conductive defect. If one cochlea is faulty it is heard more loudly in the intact cochlea.

7.30 Rinne's Test

Hold the vibrating fork (any frequency in range 64 to 512 Hz) with its base firmly on the mastoid process. You will hear the note by bone conduction. It will decline in loudness as the output of the fork declines with time. When it is just inaudible take it off the mastoid process and hold the prongs near the outer ear. Can you hear it now? If you can, it means your air conduction hearing is more sensitive than your bone conduction, this is normal and is called a positive Rinne. If you cannot, repeat the test the other way round, i.e. listen by air conduction until you just cannot hear the note and then put the base of the fork on the mastoid process. If you can now hear it, this suggests conductive hearing loss and is called a negative Rinne. Hearing loss associated with a positive Rinne suggests perceptive deafness.

7.31 Weber's Test

Rest the base of a vibrating tuning fork on the forehead in the midline or on the top of the head. Describe your sensations. Is the sound louder in one ear than the other? Does it seem to be localized on one side?

7.32 If in the preceding test the sound was heard equally loudly in both ears, you can imitate conductive deafness in one ear by blocking it with a finger tip. Repeat the test: does the sound seem equally loud in each ear now? Place the base of the vibrating fork on the mastoid process, listen to the sound, now block your ear with a finger tip. Does the loudness of the sound increase?

7.33 Masking

Ask the subject to read from a book. After a few sentences make a rattling noise, using a tin box containing pieces of lead, near his ear. The intensity of the voice will be raised. This would not occur in a deaf person. This test is used to detect malingering.

The same result can be achieved more elegantly and with less disturbance to the rest of the class if the subject wears a pair of headphones connected to a buzzer or some device which produces a noise in the phones.

Localization of a Sound Source

The brain uses three clues in localizing a sound source:

(a) the relative loudness of the sound heard by the two ears
(b) the time relationship of the arrival of the sound at the two ears
(c) transformations of the quality of the sound produced by the angle of the pinna to the sound wave front.

Estimates based on (a) and (b) need binaural hearing, the reduced localizing power of individuals deaf in one ear is accounted for by factor (c).

7.34 The subject closes his eyes. The observer makes clicking noises with forceps from various positions around the subject's head. The subject attempts to localize the sound source. Block one auditory meatus with wet cotton wool or the tip of the subject's finger. Repeat the test and see if the subject is as accurate as before.

7·35 The contributions of loudness differences in the two ears and time-delays in the arrival of sounds at the two ears can be studied precisely by routing signals generated by a stimulator separately to the two ears through headphones. The general arrangement is shown in Figure 7.6. The instructions for displaying the output of the double-pulse stimulator on the double-beam cathode ray oscilloscope are given in detail in paragraph 1.10. Set the controls of the instruments as follows:

Stimulator

Frequency 10/sec
C1 delay range > 1 mS, fine control—midway
C2 delay range > 1 mS, fine control—midway
C1 width midway
C2 width midway
C1 intensity range 0 to 18 v, fine control 2
C2 intensity range 0 to 18 v, fine control 2
Output switch—C1 (i.e. the signals arrive separately at outputs C1 and C2)
On/Off switch—On.

CRO

Trigger—external
Stability—just less than free running
Level—set so that trigger pulse initiates timebase sweep [for further details of this adjustment see paragraph 1.10]

Time base velocity	1 mS/cm
Y1 amp sensitivity	0·2 v/cm D.C.
Y2 amp sensitivity	0·2 v/cm D.C.

Put on the headphones. Clicks at 10/sec should be heard. Turn intensity C1 and C2 alternately to zero and decide which channel is which. C1 is the left ear channel and is connected to the upper Y1 beam. Adjust fine delays on C1 and C2 so that the square waves are in midscreen, one below the other. Adjust C2 intensity so that the height of the square wave is the same as the two beams. Adjust C2 width to give the same width as C1. You should now be hearing simultaneous clicks of equal loudness and pulse-width in each ear. With the C2 FINE DELAY CONTROL advance the C2 pulse so that it arrives in the right ear 1 msec before the C1 pulse arrives in the left ear. Now move the control to retard it by 1 msec. Repeat several times with eyes closed. The apparent position of the sound source should move around the subject's head. Assuming a sound velocity of 34 cm per msec and the distance between the two pinnae as 25 cm, delays and advances of approximately $\frac{3}{4}$ msec cover the possible events as the sound source moves round the subject's head. Now arrange the stimuli for simultaneous arrival but vary the relative loudness in each ear. Begin with no sound in the left ear and comfortably loud in the right. Gradually increase the left ear intensity until it is equal to the right ear intensity. Now reduce the intensity of the right ear stimulus to zero. Describe your sensations.

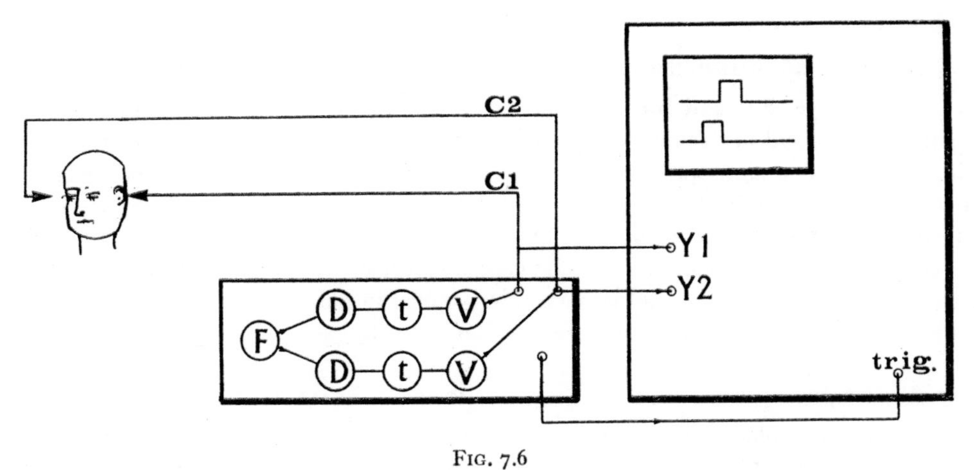

FIG. 7.6

See para. 7.35.

Arrange the stimuli so that the left ear has a msec advance and a signal of twice the amplitude of the right. Now decrease the advance and amplitude of left ear signal until at zero advance the amplitudes are the same. Then put in a gradually increasing delay in the left ear signal and simultaneously decrease the amplitude to one half of the right channel signal. Describe your sensations.

VISION

7.36 Mechanical Stimulation of the Eye

Close the eyelids and press on the eyeball. Describe the sensation and note its position in the field of vision for comparison with the position of the part of the retina stimulated. This visual sensation is called a Pressure Phosphene and is a good example of the general principle that if a sense ending is stimulated, the central nervous system interprets the afferent nerve impulses in terms of the sensory modality associated with the sense organ; even though, as in this case, the sense ending has been stimulated by an inappropriate stimulus. Another example is given in paragraph 7.25 where electrical stimulation of the tongue produces a sensation of taste.

7.37 Blind Spot

1. Make two black circles about 3 mm diameter and 10 cm apart. Hold the paper up in front of the right eye at arm's length; close the left eye. Fix the right eye on the left-hand mark. Bring the paper slowly towards the face. The right-hand mark will disappear and then reappear as the paper is brought nearer.

2. Rest the subject's chin on a support (e.g. a book) about 25 cm above the bench. He gazes steadily (left eye closed) at a small cross in the centre of a piece of white paper below his right eye. The observer prepares a long strip of white paper with a large (3 mm) black dot at one end; holding the other end he makes the black dot travel over the paper from the right towards the cross. When the dot becomes invisible to the subject and when it reappears, the observer makes a mark through the dot on to the paper below. This procedure is repeated for vertical and oblique movements of the dot across the blind area. Finally the points are joined up to make an outline of the projected image of the blind spot. The largest diameter is measured and the actual size of the blind spot is calculated by the method of similar triangles, given that the distance between the nodal point of the eye and the retina is 15 mm. Calculate also the distance of the blind spot from the visual axis—i.e. the fovea centralis. Approximately how many times the diameter of the blind spot is this?

7.38 Near Point

Hold an open book in front of the eye and bring it nearer until the print can no longer be seen clearly. Measure the distance to the eye. Do this again without spectacles if these are worn. Repeat this experiment at home on people of various ages. What influence does age have on the near point?

7.39 Focusing

Hold a pencil between one eye and the corner of the room. Keep the other eye closed. Attempt to focus both the corner of the room and the pencil at the same time. Illustrate the result by means of diagrams and discuss the problem.

7.40 Visual Acuity in Various Parts of the Field of Vision

A large semicircle has been drawn on the surface of the bench. The subject places one eye just above the bench level at the centre of the semicircle and looks steadily at a fixed point, a thin short rod placed on the semicircle directly opposite. The other eye is kept closed. Angles of divergence from the visual axis are marked out on the semicircle. The test object consists of a card showing an incomplete black circle on a white ground. A number of cards each with a different-sized gap are provided. The gap is a detail which can be detected only if its image falls on a part of the retina with adequate visual acuity.

The card, mounted on a small block of wood, is moved out from the fixation point until the subject just fails to see the gap. The angle is noted. The card is now brought in from the periphery and the angle at which the gap becomes visible is noted. Discover the connection between the size of the detail and the angle at which it can be seen.

7.41 Field of Vision

The charting of the field of vision is often of great practical value in the localization of lesions in the brain.

1. For clinical purposes there is a simple procedure which can detect gross changes. The subject stands with his back to the light and facing the examiner at a distance of about 1 m. Each eye must be examined separately; while one is being tested the other should be closed. The examiner closes the eye opposite the subject's closed eye. The subject should look fixedly at the examiner's open eye, while the latter, holding his hand midway between himself and the subject, moves the outstretched forefinger from the periphery towards the centre of the visual field. The subject is asked to say when he sees the movement of the finger. Both the examiner and the subject should see the movement of the finger at the same moment provided that in both the field of vision is normal. The movements of the hand are repeated in different meridians of the field until it has all been explored. Thus the examiner's field is compared with the subject's, and as the examiner is constantly watching the subject's eye any wandering from the point of fixation is quickly observed and corrected.

2. A more accurate examination can be made by the perimeter$_{14}$. This consists of a metal arc rotatable about its centre so as to describe a hemisphere, mounted on a stand in front of a large black disc. At the centre of the arc there is a small white disc or a small plane mirror used as a point of fixation. At the opposite end of the instrument is an adjustable chin-rest on a pillar. The subject places his chin on this and the height is adjusted till the eye to be examined is on the same level as the fixation point; the adjustment is correct when the subject can see his own eye in the central plane mirror. The subject should sit with his back to the light and close the other eye or have it covered with a shade.

The mode of writing out the results of a perimetric examination depends on the type of instrument. With the simplest models, the inclination of the arc is read from a dial on the stand and the position of the test object is read from a scale engraved on the arc. Transfer these readings manually to appropriate points on the perimeter chart.

With automatic perimeters there is a mechanism which permits the position of the test object relative to the eye to be marked at will on the perimeter chart. Examine the perimeter you are to use and ask the help of a demonstrator, if necessary.

Use the white target and take it to the end of the quadrant; bring the target inwards until it just comes within the field of vision of the eye of the subject who is gazing steadily at the fixation point. Record this position. Repeat this procedure with different inclinations of the quadrant. Finally join up the dots on the chart to get the boundary of the field of vision and compare it with the field already printed on the chart. Try to account for the peculiar shape of the field.

The perimeter may, of course, be used to map out the blind spot by bringing the target along the quadrant in several meridians close to the horizontal on the temporal side of the field.

3. If a perimeter is not available the horizontal and vertical limits of the field of vision can be mapped out by chalking semicircles on the bench and on a vertical blackboard. The subject places his eye at the centre of the semicircle (which is placed at the edge of the bench or blackboard) and gazes steadily at a piece of chalk on the circumference opposite him.

The observer uses a chalked white disc on a long black handle to keep his hands out of the subject's field of vision. Bring the disc from the periphery of the field of vision along the semicircle towards the fixation point. When the subject sees the white spot a chalk mark is made. The procedure is repeated from the other end of the semicircle.

7.42 The Fundus Oculi. This work is done in a dark room.

The use of the ophthalmoscope will be demonstrated. Afterwards, practise viewing the fundus of the model eyes to determine whether the image is erect or inverted. Then examine the human eye.

Both subject and observer must relax their accommodation by 'looking into the distance' in spite of the fact that the object looked at is only a few inches away. Both must remove spectacles if these are worn. Use the right eye to look at the subject's right eye and the left eye to look at his left eye. Rotate the knurled disc which controls the chain of lenses until there is no lens opposite the eye-hole or until the algebraic sum of the observer's and subject's spectacle lenses has been set up. Ask the subject to keep his eyes fixed on some suitable object or fixation point on the opposite wall. Then beginning at arms' length from the subject and just clear of his line of sight look through the aperture of the ophthalmoscope whilst steadying it against the bridge of your nose and shine the beam on to the subject's eye. You will now see a glow from the pupil. This may be red (red reflex) or white (if you are in line with the optic disc). If necessary rotate the ophthalmoscope slightly. Gradually decrease the distance and you will see more detail. Note the pale optic disc and the retinal vessels emerging from its centre; the arteries have a rather brighter red colour and reflect the light more strongly than the veins. To keep his eye steady the subject must avoid looking at the ophthalmoscope and gaze steadily with the other eye at the distant fixation point. Remember that you are simply using the lens of the eye as a magnifying glass to examine the fundus and that the essential function of the ophthalmoscope is to illuminate the interior of the eyeball. It may help at first to use a weak minus (i.e. concave) lens at the eye-hole of the ophthalmoscope. This has the effect of neutralizing your accommodation which you may have difficulty at first in relaxing. After a little practice this difficulty disappears.

Sketch the appearance of the fundus, asking the subject to look up, down and to the side so that as much of the fundus as possible is seen.

If two members of each section will volunteer to allow homatropine drops (1 per cent.) to be put into one eye about half an hour before the examination this will allow every one to have a good view of the fundus. The effect can be counteracted afterwards by instilling 1 per cent. eserine. What are the actions of atropine or homatropine and of eserine? Repeat experiments 7.38 and 8.4 on the volunteers.

Use a plus 20 lens in the ophthalmoscope—the corneal surface is now easily seen. A plus 12 will enable the anterior surface and a plus 8 the posterior surface of the lens to be focused.

7.43 Colour Mixing

Accurate colour mixing can be achieved with beams of light more satisfactorily than with pigments. The apparatus consists of three spotlights fitted with filters which transmit red, green and blue light respectively; the filters are those of the tricolour photographic series. A fourth spotlight supplies white light when required. The lights can be switched on individually and controlled in intensity by means of the variable resistances on the front panel; the beams are aligned to converge on a small opal glass plate where mixing takes place.

Switch on the three coloured lamps (neglecting white for the time being) and note the colours produced by variation of the red, green and blue components. It is possible to produce a satisfactory white, and to match almost any colour, although the adjustment is often quite critical. Pastel shades, which contain a proportion of white in their make-up, can be demonstrated by diluting any selected colour or mixture with varying intensities of white light from the fourth spotlight. The resistances have been adjusted so that the position of the slider corresponds roughly to the proportion of the primary colour in use; a colour mixture can therefore be specified approximately in terms of the proportions of its component primaries.

7.44 Tests for Colour Blindness

Read the instructions for Ishihara's Test Cards. These consist of a series of plates, all with splashes of colour in the same irregular pattern, but the colours differ from plate to plate. In the pattern letters are formed in splashes of one colour on a background of splashes of a different colour. The subject is asked to name the letters or figures so picked out.

7.45 Edridge-Green's colour lantern is used to test engine-drivers, aviators and sailors. It is so arranged that different colours at varying apertures can be shown—these represent railway signals and navigation lights at varying distances. There is also a series of modifying glasses by which the effect of fog, etc., on the colour of signals can be imitated.

7.46 Field of Vision for Colours

Estimate the field of vision for white, green, red, yellow and blue by the method already described (7.41(2)), using coloured discs or lights. The sensation must be more than a mere awareness of something in the visual field. Move the disc in from the periphery until the subject is certain he can distinguish the colour.

7.47 After-Images

These three experiments are performed in a dark room.

1. Gaze fixedly at a bright circle of light (opal electric bulb) for 20 seconds. Turn off the light and look at the black surface of the wall. A positive after-image, owing to the absence of a second stimulus, will be seen.

2. Again gaze fixedly at the light for 20 seconds, then quickly look at the centre of a large white area. Owing to the second stimulus (the white area) the after-image will now be negative and will appear as a dark area on the white ground. Note that the after-image is always of the same shape and size as the original stimulus.

3. Look steadily at the piece of red glass which is illuminated from behind. Then look at a white area. Again there will be a negative after-image, in this case green, which is the complementary colour. Repeat with blue and yellow.

7.48 Myopia and Hypermetropia

Kühne's artificial 'eye' consists of a rectangular water-filled tank with a convex circular glass window at one end to represent the cornea. Within the tank, immersed in the water are a diaphragm to represent the iris, a bi-convex lens, and a ground-glass screen to represent the retina. A few drops of fluorescein solution are added to the water in the tank to make the rays of light visible. Place a luminous object such as a candle or a clear car lamp (say 12 V, 36 W) in front of the eye and move the screen till the image is sharply focused. This will represent the condition of emmetropia. Now push the screen forward nearer the lens. The anteroposterior measurement of the eye is now too short and the image of the object is poor. If now a suitable convex lens is placed in front of the eye the image is once more in focus. The same result could be brought about in the living eye by accommodation—the lens then becomes more convex and so a good image is formed. This shortness of the eyeball is the commonest cause of hypermetropia or longsightedness. If the object, the candle or the lamp, is brought too near the eye, then accommodation is not sufficient to bring the image to a sharp focus on the retina. That is to say, the near point may be at a considerable distance from the eye and objects close at hand cannot be seen clearly unless, as in this experiment, a convex lens is put in front of the eye.

Now push back the screen so that the eye is too long, the condition usually found in myopia. The image will be poor unless the candle is brought nearer to the eye (hence the usual name shortsightedness), or a concave or minus lens is placed in front of the eye.

Return to the position of the emmetropic eye. Now push the candle nearer the eye—the image deteriorates. This is very nearly the state of affairs in presbyopia where the lens has lost its elasticity and accommodation cannot occur. What kind of lens must be given to an old person who finds he has to hold his book at an uncomfortable distance for reading?

The artificial eye is supplied with a metal plate with a circular aperture which is placed in front of the lens. Removal of this plate will have the same effect as full dilation of the pupil. Focus the image on the screen and then remove the plate. Note the effect on the sharpness and brightness of the image.

Return to the position of the emmetropic eye. Place a cylindrical lens in front of the eye. This imitates a toric defect in the cornea and the image is blurred in one axis. Rotate the cylindrical lens and note the effect on the image.

7.49 Visual Acuity

Visual acuity is tested by Snellen's types. A card on which are printed rows of letters of different sizes is viewed in a good light from a distance of exactly 6 metres. Below each row is a number indicating the distance in metres at which that row can be read by a person with normal vision. This is the distance at which the width of the black strokes forming the letters subtends at the eye an angle of one minute and the whole letter subtends an angle of five minutes.

The subject is placed at 6 metres from the test type and uses one eye at a time. The observer indicates the letters with a pointer and the subject reads them off. The observer notes the number below the last row that can be read correctly. The visual acuity V is expressed as 6 over this number; normal vision is therefore 6/6.

Lessened visual acuity may be due to (1) defect in the nervous apparatus—i.e. retina, nerve or brain, (2) defect in the media of the eye—e.g. cornea or lens, (3) ametropia, or to a combination of these. To show that lessened visual acuity is due simply to (3) and not to something more serious, the principles of the pin-hole camera can be applied. Make a pin hole in a piece of black paper. If you have bad visual acuity remove your spectacles; if you have good acuity put a lens in front of the eye (spectacles borrowed from a colleague) to disturb it. Look through the pin hole held close to the eye—brightly lit objects are then clearly seen in the absence of defects (1) and (2) above.

7.50 Astigmatism—Astigmatic Fan

The set of radiating lines on the card is known as an astigmatic fan. Cover one eye and look at the fan with the other eye. Do all the lines appear equally black? If some are black and others grey or blurred, then astigmatism is present. Note which

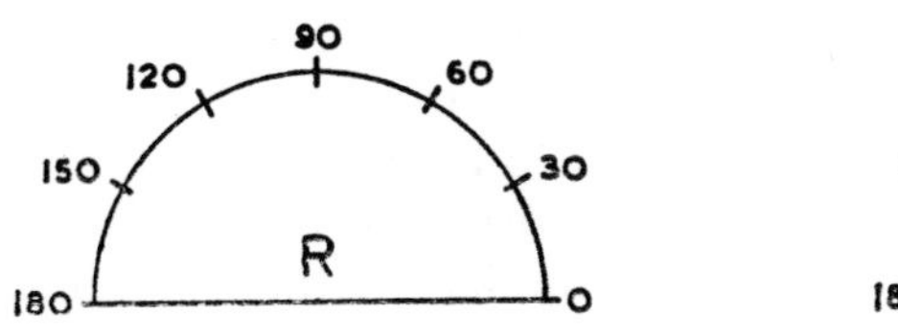

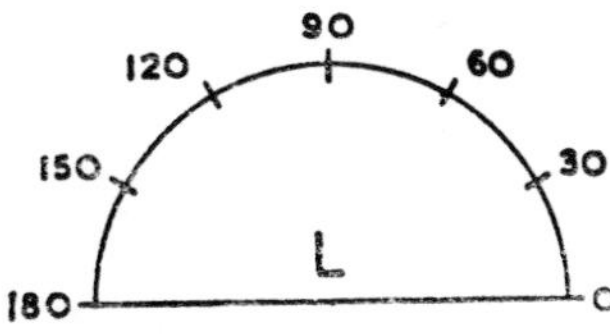

FIG. 7.7

line is seen most clearly. Repeat with the other eye. The line at right angles to this line is usually the most blurred and corresponds to the axis of astigmatism. The axis is usually indicated by drawing a stroke through the accompanying chart (Fig. 7.6). It is easier as a rule to tell which line is most clearly seen rather than which line appears most blurred. If glasses are worn carry out the test with and without glasses.

In astigmatism the image of a point source is not a point, hence the name (see 7.48). In regular astigmatism such as is produced by combining a spherical lens and a cylindrical lens there are two linear foci at right angles to each other. If one of these foci coincides with the retina, then the image of a straight line parallel to this linear focus will consist of a series of overlapping short straight lines side by side, i.e. a rectangular patch with a width equal to the length of the linear focus. Such a straight line will therefore appear blurred.

7.51 Placido's disc

Stand facing a good light and look through the central aperture of Placido's disc at the subject's eye. A convex lens in the aperture magnifies the reflexion from the cornea of the black and white concentric circles of the disc. If the cornea is spherical the rings are regular and truly circular. Regular astigmatism causes the rings to be oval in shape though still concentric. Irregular astigmatism, conical cornea, etc., distort the rings in various ways.

7.52 Cover Test

Ask the subject to look at your right index finger held about 0·5 m from his eyes. While the subject is gazing steadily at your finger cover his right eye with your left hand so that the left eye only can see your finger. Move your left hand over to cover the left eye. If the right eye moves inward or outward at the moment of taking up fixation a squint is present; if it remains motionless there is no squint.

Double Vision

7.53 Normally the external eye muscles orientate the optical axes of the two eyes so that the two images of the part of the external world under scrutiny fall on corresponding points of the two retinae. If, however, the action of the extraocular muscles of one eye is disturbed by an externally applied force the image on one retina will be displaced and double vision will result. To test this, look at an object on the bench about 1 m away, press gently with a finger on the upper eyelid and observe that two images of the object separate themselves. On the other hand it is quite normal for the images of objects not under scrutiny to be double; hold a pencil at about 30 cm from your face and look at an object at the far end of the room, the pencil will appear double. Now focus on the pencil and objects in the background will appear double. Draw diagrams to illustrate this point.

7·54 A latent defect of neuromuscular co-ordination can be made apparent by any method which makes the two fields of vision so dissimilar that no fusion is called for. The Maddox rod consists of a red glass with a corrugated surface. Put on the test frame and put the glass in front of one eye while looking at a bright light. The covered eye will see a bright red streak, the direction of which depends on the axis of the corrugations. Does the streak pass through the light or to one side? What does this indicate?

7·55 Make a series of figures from 1 to 10 across the pages of your book with an arrow about 1 cm below the 5 and pointing up to it; look at this diagram at a distance of 25 cm through the spectacles, which have discs of metal instead of lenses. The discs have narrow horizontal slits which are continued into narrow tunnels. Move the head about until by testing (by closing the eyes alternately) only the row of figures is seen through one slit and only the arrow through the other slit. Read off the position of the arrow with both eyes open; afterwards enter a dotted arrow there. Does this correspond to the actual position of the arrow? If not, what does this indicate?

The Maddox wing[14] has two apertures in the eyepiece through which a screen carrying a row of figures and an arrow is viewed. Between the eyepiece and the screen are two wings so that the left eye sees only the row of figures and the right eye only the arrow. With both eyes open note the position of the arrow relative to the row of figures. Is this the actual position of the arrow?

7·56 Binocular Field of Vision

Draw as large a semicircle as possible on the top of the bench. Put the bridge of the nose at the centre and place a piece of white chalk on the circle opposite to act as the fixation point. Using the method previously described (7·41(3)) find the field of vision of the right eye, then the left eye and then of both together; keep the head in the same position throughout. Measure the angles and make a diagram. What advantage of binocular vision is shown by this experiment? Bring the fixation point (i.e. the piece of chalk) nearer the nose so that it is about 3 cm farther away than the near point. Make a circle of this radius and proceed as before. Compare with the previous result.

7.57 Stereoscopic Vision

Thread a needle; note how you do it and how long it takes. Close one eye and repeat the experiment. How does this affect your performance? You must not allow your hands to touch one another during this test; otherwise tactile clues will reduce the effect of closing one eye. Note if there is any difference in the ease with which the experiment is completed when one eye alone is used if the hand holding the thread and the hand holding the needle approach one another (a) along the visual axis and (b) at right angles to the visual axis.

7.58 Measurement of Accuracy of Stereoscopic Vision

The apparatus consists of three silver-painted vertical rods which are observed through a hole in a screen; the hole is placed a few inches from the rods and arranged so that only the centre parts of the rods are visible. The two outer rods are fixed, and the centre one can be moved along a scale to lie before or behind the line of the others by a measured amount.

The subject sits about 1·5 m from the screen and the observer moves the middle rod until it appears to the subject to lie in the same plane as the others; the error can be read from the scale. Take six readings, and work out the average error; convert this into a percentage of the distance of the rods from the subject. Repeat the experiment using one eye only and compare the effectiveness of monocular and stereoscopic vision.

7.59 Hold one of the wooden blocks provided about 25 cm in front of the eyes. Describe its appearance as seen by the right eye alone and then by the left eye alone. Show that two slightly different views of an object can be fused to give an impression of solidity when they are combined with the help of a sterescope—either Wheatstone's or Brewster's. Draw sketches to indicate the paths of the light rays when using these stereoscopes.

7.60 An X-ray photograph is really a shadow photograph and gives no information about the relative position of the various structures. Before the operative removal of foreign bodies, such as bullets, accurate localization is possible by combining in a stereoscope two skiagrams taken from slightly different positions. A modified Wheatstone apparatus for transillumination of X-ray negatives is available. Move the films about in their holders until a stereoscopic effect is obtained. Make a diagram of the apparatus and explain the appearances.

CHAPTER EIGHT

THE NERVOUS SYSTEM

Reflexes in the Human Subject

If a normal reflex is obtained it indicates that the reflex path—sense organ or receptor, ingoing nerve, connections in central nervous system, outgoing nerve, muscle or effector—is intact. Reflexes may be exaggerated, diminished or altered. It is important to know the normal reflex in order that any departure from it may be readily recognized.

Tests of reflex function in human subjects raise particular problems. Firstly, many reflex activities, normally controlled by the lower levels of the central nervous system, can be extensively modified by the activity of higher centres. Thus when spinal reflexes are being tested the subject must be kept at ease, both mentally and physically, and his attention distracted from the part under investigation.

Secondly, it is difficult to test a single sense; we normally receive information from several senses simultaneously. The experimenter must devise and conduct tests where the subject cannot obtain information from other sources, i.e. where the subject cannot 'cheat', either at a conscious or a subconscious level.

Lastly, where the subject's co-operation is required, he must be given precise, simple instructions, couched in language which he can understand. There is no problem here if the subject has had biological training but few people in the world at large are familiar with physiological terminology. It would be pointless to tell the average person to 'Focus on a near object'; he must be told, 'Look at my finger'.

8.1 Cutaneous or Superficial Reflexes

These cannot be tested in the class but must be tried out at home.

1. *Plantar Reflex.* Scratch the sole of the foot near the inner side. All the toes become plantar flexed. If the hallux is pointed upwards and the other toes fanned, this is described as the Babinski or the extensor type of plantar response, and in the waking adult subject indicates damage to the pyramidal fibres. It is, however, the normal response of a child in the first year of life when these fibres are still myelinating; it can also be elicited in adults who are deeply asleep.

2. *Abdominal Reflex.* Stroke the skin of the abdominal wall. The reflex contraction of the abdominal muscles will pull the umbilicus to the side stroked.

3. *Cremasteric Reflex.* Stroke the inner side of the thigh. There will be a contraction of the cremaster muscle.

8.2 Tendon or Deep Reflexes

1. *Knee Jerk.* The subject must be seated in a chair or on a stool of moderate height, with legs dangling or with one leg crossed over the other. He must not be in a strained position and if possible his attention should be diverted from the experiment. With the tendon hammer hit the patellar tendon just below the patella. The extensor muscles will contract and the foot will be kicked forward. If no reflex is obtained ask the subject to hook his fingers together and to pull them apart. Try again to elicit the reflex. This is a reinforcement manoeuvre.

2. *Achilles Jerk.* The subject kneels on a chair so that both legs are supported but the feet hang free. Hit the tendo achillis smartly with the tendon hammer and note the movement of the foot caused by the contraction of the gastrocnemius muscle.

3. *Biceps Jerk.* Ask the subject to let his arm lie slackly, his forearm supported on your left forearm while you steady his elbow on the palm of your left hand. Place your left thumb on the biceps tendon and hit your thumb with the hammer. The biceps will contract and make the tendon tense under your finger.

4. *Triceps Jerk.* The subject's arm should lie slackly. Support his upper arm in a horizontal position with your left hand and let his forearm hang vertically down. Tap the triceps tendon just above the elbow joint and note the movement of the arm.

8.3 More can be learnt about the time relations of the knee jerk by using electrical methods to record the events in the quadriceps muscle. The apparatus is arranged as in Figure 8.1. The tendon hammer A has a switch and probe mounted in it so that a trigger circuit to the time base of the oscilloscope is closed when the hammer strikes the knee. This circuit sets the luminous spot moving along the X axis of the cathode-ray tube face at a steady speed. This speed is accurately known and is determined by the setting of the time-base controls.

The sequence of events during a knee jerk is as follows. The tendon hammer strikes the tendon and causes a small displacement. The displacement is transmitted to the quadriceps muscle and causes an abrupt but small elongation of the muscle. The elongation is signalled as a modification of the discharge from muscle spindles within the muscle and reflexly causes a short-lasting increase in the discharge of motor impulses to the extrafusal fibres of the quadriceps. This increase causes the jerk. Silver-foil electrodes fixed to the skin over the quadriceps are connected via the oscilloscope amplifier to the Y plates. The synchronized muscle action potentials of the reflex jerk appear as a deflexion on the Y axis. Since the spot speed along the X axis is known in mm per msec, a measurement of the distance between the start of the time-base and the jerk deflexion gives the reflex time. Note that the amplitude and form of the

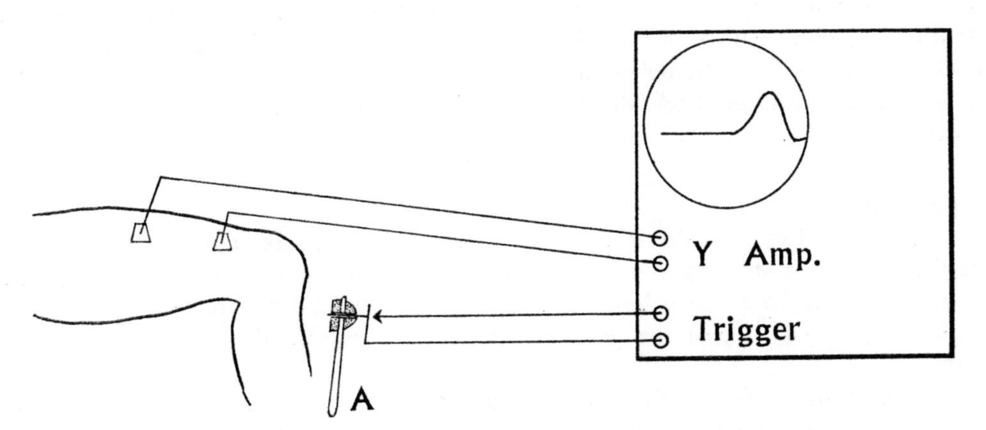

FIG. 8.1

jerk electromyogram can be greatly altered by reinforcement manoeuvres (see 8.2,1). Measure the nerve pathways from the quadriceps muscle to the lumbar spinal cord and back. Neglect the synaptic delays and calculate the average speed of conduction of the message to the cord and back. Read up in your textbook about the types of fibre which innervate the muscle spindle and the extrafusal skeletal muscle fibre.

8.4 Eye Reflexes

1. The subject faces the window. Look at his eyes and note the iris (the coloured part) and the size of the pupil. With your two hands cover both eyes for 10 seconds. Instruct the subject to keep looking at a distant object. Then remove one hand and observe the pupil. Repeat the procedure for the other eye. This is the direct light reflex.

2. Keep the subject in the same position and cover both eyes again. Stand to one side and raise one hand slightly to observe the pupil, but not so much as to let light from the window into this eye. Remove the hand entirely from the other eye while continuing to observe the pupil of the first eye. This is the consensual light reflex.

3. Reaction on accommodation or convergence. Turn the subject so that he is facing a dully illuminated wall. Ask him to look at the wall and then at your finger held about 15 cm from his eyes slightly above eye level (to keep the upper lids raised). Observe the change in the pupils.

4. The corneal and conjunctival reflexes are really superficial reflexes. Gently touch the periphery of the cornea, or the sclerotic, with the corner of a clean handkerchief or a piece of clean paper. The eyelids close immediately; if the stimulus is more intense lacrimation will occur.

These are the reflexes usually tested in the routine examination of a patient. Describe the normal reflexes and indicate which part of the central nervous system is involved in each.

8.5 Reflex Action in the Frog

1. Study the behaviour of the intact frog and its response to stimulation of various kinds. Note how it rights itself when placed on its back. Observe respiration.

2. Remove the cerebral hemispheres by cutting off the head of the frog forward of a line running just caudal to the eyes. After a few minutes examine the frog's behaviour once more. Do respiratory movements continue? If very active, this preparation may swim in a basin of water. Place the frog on a rough board or on the frog board, tilt the board and the frog will slowly move up.

3. Remove a further zone of the nervous system by cutting off the head forward of a line passing behind the tympanic membranes. After a few minutes, observe the attitude and behaviour of the preparation. Apply a 5 mm square of blotting paper soaked in 10 per cent. acetic acid to the frog's leg. Note that the remaining part of the nervous system is able to produce co-ordinated movements which may dislodge the irritant.

4. Pinch one toe with forceps and note the reflex response. Vary the strength of the stimulation and see if you can detect variations in the duration and extent of the response.

5. See if there is any difference in response if the toe is pinched first with the leg in a flexed posture and then in an extended posture. Does it show a thrust response?

6. Turck's time of reflex action. Have ready three jars: (*a*), plain water; (*b*), weak acid (0·5 per cent. sulphuric acid); (*c*), stronger acid (1 per cent. sulphuric acid). Hold jar (*b*) so that the acid covers the frog's legs to just over the knees. Measure the time till reflex action is elicited then wash off acid with (*a*). Similarly try the effect of (*c*), measuring the time as before. Wash off with (*a*).

7. Make an estimate of the tone in the limb muscles of the frog by moving the joints. Now destroy the spinal cord by pushing a wire down the vertebral column (pithing). Re-examine the tone of the limb muscles; they should now be quite flaccid as a result of destruction of the spinal nerve centres. Test for reflexes, these also should have disappeared.

8.6 The Human Electromyogram during Graded Voluntary Contraction

Apply metal foil surface electrodes to the skin over the biceps muscle. The electrodes are connected to a differential-input amplifier and thence to a loudspeaker. An ink-writing galvanometer or cathode-ray oscilloscope can be connected in parallel with the loudspeaker if visual indication of the electrical activity of the muscle is desired. The subject should be earthed with an electrode applied to the neck or any other convenient position that does not overlie active muscle. The subject stands with his arm hanging down. In this position there should be no sound from the loudspeaker. Now the subject bends his elbow until the forearm is horizontal. Note the faint signals from the loudspeaker. The experimenter now places weights of increasing size, e.g. 0·5 kg., 1 kg., 2 kg., into the hand of the subject and tells him to maintain the forearm horizontal. Note the graded increases in electromyographic activity.

8.7 Motoneurone Discharge Rates During Voluntary Movement

Action potentials in motor nerve fibres innervating skeletal muscle produce muscle action potentials of the same frequency in each of the muscle cells comprising the motor unit. Thus if a pick-up electrode is arranged to detect the activity in a very small group of muscle fibres, ideally all part of the same motor unit, information can be gained of the activity of the motoneurone controlling the unit. There are a few favourable situations, e.g. the tongue, around the lip margins and other parts of the facial musculature where the activity of single motor units can be picked up with sharp silver wire electrodes, separation 2 mm, pressed into the skin surface, without penetration. The wires are connected to a high-gain differential input amplifier and oscilloscope, maximum overall sensitivity 100 μV per cm. The subject should be connected to earth by an ECG-type plate electrode to reduce mains interference. A soundmonitor with loudspeaker is connected in parallel with the oscilloscope.

The subject should explore the lip margins and tongue surface with the electrodes, whilst making slight voluntary contractions of the muscles beneath the skin, until a position is found where the activity of a single motor unit can be seen in isolation on the cathode-ray oscilloscope. Now study the frequency of discharge associated with voluntary contraction.

8.8 Cold Pressor Test

The subject sits comfortably and readings of systolic and diastolic blood pressure are taken at intervals of one minute until the readings are steady. Usually the blood pressure stabilizes only after several minutes; this stable value is the one which is useful clinically. The initial reading may be quite misleading.

The free hand of the subject is then submerged in ice-cold water up to the wrist for one minute. Readings of SBP and DBP are taken after 30 seconds and 60 seconds of immersion. Further readings are taken in each of the following 5 minutes, or until the blood pressure falls to a stable value. Plot the systolic blood pressure, diastolic blood pressure and pulse pressure on the same graph.

In young people this experiment produces a rise in SBP of the order of 20 to 30 mm Hg.; the initial value is regained a few minutes after the hand is removed from the cold water. Describe the components of this reflex.

8.9 Electrical Resistance of the Skin

The resistance offered to the passage of an electric current by the body is determined largely by the outer layers of the skin which are poor conductors. The resistance is greatly modified by changes in skin blood flow and sweat gland activity. Changes in resistance reflect changes in the activity of the autonomic nervous system which controls the superficial blood vessels and sweat glands.

The principle of the measurement is simple. A battery (9V) passes current through a meter (full scale deflexion 50 μA) and through the tissues, the amount of current being proportional to the resistance according to Ohm's Law. If the battery voltage is fixed the current meter can be calibrated in ohms. Note that the meter is very delicate and will be damaged if the electrodes are touched together when the battery is in circuit. A push button is included in the circuit to isolate the battery while the electrodes are being applied.

The test electrodes are metal plates which are placed across the hand, one on the palm and one on the dorsum, and retained there at a constant pressure by means of an insulated clip. When the electrodes are applied the hand and the forearm should rest comfortably on the bench. When making a measurement press the push button just long enough to read the meter deflection. The current produces slight electrolysis in the skin which changes its resistance.

Measure the resistance between the palm and the dorsum of the hand (a) in the natural condition; (b) after thorough washing with soap and water and drying; (c) after immersion of the hand in hot water for a few minutes, and drying; (d) after swabbing back and front with saline and leaving wet. Record your results and account for them. What bearing have they on the liability of a person to serious accidental electric shock?

Try to demonstrate sweat gland activity in response to emotion by making measurements of skin resistance. Wash the electrodes and dry them thoroughly. Clip them across the hand and note the reading on the meter. Startle the subject or cause pain by a pin prick or by pulling a hair with forceps. There is often a sudden and quite considerable reduction of resistance as autonomic activity provokes sweat gland secretion. A similar but rather more sensitive device, popularly known as the 'lie detector', uses the same principle, namely that emotion (fear or embarrassment) causes in most persons unconscious sudomotor activity. It is obvious that the psychic reason for the emotion is in no way indicated by this device.

Try to demonstrate that heating of one hand causes reflex sweating in the opposite hand. Wash and dry one hand and clip the electrodes across it. Measure the resistance. Place the *other hand* in water as hot as can be borne and keep it there. Measure the resistance of the hand at 2-minute intervals. As a rule reflex sweating and vasodilation occur within a few minutes but there is a large individual variation.

Reaction Time

When a subject is asked to perform an action, however co-operative and alert he may be, there is an interval of time between the signal to begin and the response by the subject. To measure this time accurately a programme is arranged with the subject; thus he may be told to press a pedal when he sees a red light or close a switch when he hears a sound. Reaction time experiments of greater complexity are easily devised. The apparatus for the simpler tests is shown in diagram form in Figure 8.2. A generator A produces pulses at a frequency of 100 per second. When the operator closes his switch B these pulses light the red lamp D and enter the counter E. The pulses pass through the subject's switch C which is spring-loaded so that there is continuity until he presses it. When the subject presses his switch in response to the lamp lighting, the supply of pulses to the counter is interrupted. The counter will now indicate on its decade tubes the number of 1/100th sec which elapsed between stimulus and response. This is the reaction time. Note the reset button on the counter which returns the decade tubes to zero. The tester should keep his switch depressed until the subject responds.

8.10 Visual Reaction Time

Carry out the test with the lamp and hand-switch 10 times and obtain the average time.

8.11 Auditory Reaction Time

The subject closes his eyes, listens for the click of switch B and immediately presses his own switch. Carry out the test 10 times and obtain the average.

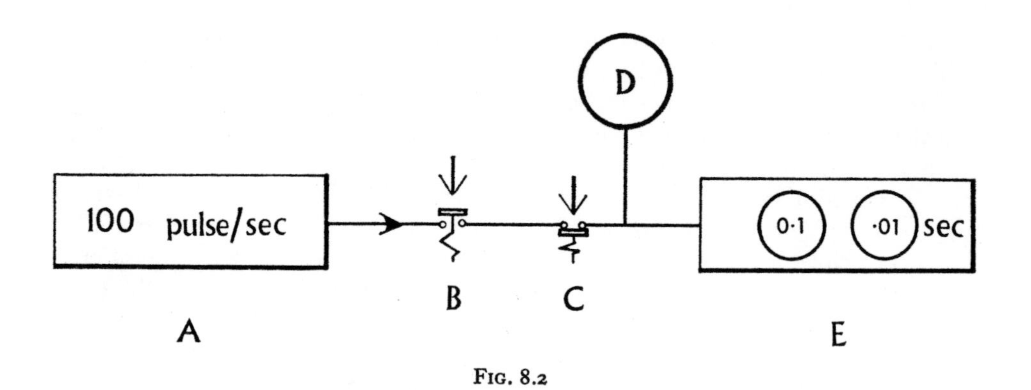

Fig. 8.2

8.12 Touch Reaction Time

The subject closes his eyes and the tester touches his hand with the contactor. Carry out the test 10 times and obtain the average. The contactor is a delicate switch mounted on a handle which closes the circuit when pressed on the skin.

8.13 Consider the elements which make up the reaction time; there is the conduction time from the sense organ to the central nervous system, the central delay, and the conduction time from the central nervous system to the effector muscles. There will be further small delays in the sense organs and neuromuscular junction and inertial delay in the effector muscles. Some estimate of the conduction time in the afferent fibres can be obtained by the following method. Measure the reaction time to a touch stimulus using a distant stimulus point, e.g. the ankle or toe; take at least 10 measurements and average. Then measure the touch reaction time using the neck; take the same number of measurements and average. With a tape measure estimate the conduction path length between the toe and the skull, similarly between the neck stimulation point and the skull. The increased time of the toe touch test can be assumed to be due to the longer distance in the afferent pathway since the other elements in the reaction time have not been changed. Subtract the neck average time from the toe average time and subtract the neck distance from the toe distance and express the result as a conduction speed of metres per sec.

8.14 Reaction time is influenced by alcohol and fatigue. This experiment is best carried out in the morning before lunch. Measure the subject's reaction time to a light signal 10 times and obtain an average. Now persuade him to drink three or more pints of strong beer as rapidly as he can without discomfort. Some judgment is needed about the amount of beer required to produce a measurable effect and it is best if a student quite unused to alcohol is not asked to carry out this experiment. Measure the reaction time every quarter of an hour for 90 minutes and express the results as a graph. Record on the graph any subjective impressions of the subject.

Reaction times are very variable according to the conditions of the experiment, see Table 8.1 for average values.

TABLE 8.1

Kind of Stimulus	Order of Reaction Time in hundredths of a sec
Sound	14
Light	18
Touch	14
Pain (without touch) . .	80

In order to give these figures a meaning for everyday affairs, take the average time for the light reaction and calculate the distance in yards travelled in that time by a motor car driven at 30 m.p.h. This will give the distance traversed before you could begin to apply the brake after receiving a warning light signal when travelling at 30 m.p.h. Note that this is only part of the distance required to bring the car to a standstill. A car travelling at 30 m.p.h. can be stopped in 30 feet from the moment of application of the brake assuming 100 per cent. braking efficiency with optimum coefficient of friction between the tyres and the road. In a real situation the car driver cannot equal the low reaction times of the laboratory because he is not prepared by a 'ready' signal. The reaction time is more likely to be about a second. Now calculate what this means in terms of distance.

8.15 Choice Reaction

Reaction time experiments can be made of almost any degree of complexity. For example, several signals may be introduced with several possible reactions. The determinations of reaction times and choice reactions can be conveniently made by using a reaction time meter. It consists of two units, linked by a multiway plug and cable. The experimenter's unit is a device for selecting and presenting a stimulus to the subject, and for measuring the time which passes before he makes the appropriate response. The subject's unit is a small box on which are mounted three coloured lamps and a buzzer—these provide the stimuli, visual and auditory—and a set of four push-button switches. The subject must press the correct button to extinguish the stimulus and stop the time-measuring device. Provision is made in the experimenter's unit for the stimuli and push-buttons to be interconnected in two different ways, one so that the relation is obvious ('direct switching'), the other less apparent ('crossed switching'). Thus a simple, or a more difficult, problem can be set.

8.16 The Electroencephalogram (EEG)

Potentials having their origin within the skull can be detected on the skin over the skull in an attenuated form: the voltages are transmitted through the conducting tissues of the head to the skin surface just as the cardiac muscle potentials are conducted to the skin of the chest and limbs. The brain waves are of small voltage (10-100 μV) and fluctuate within a restricted range of frequency (1-100 c/s.) The ink-writing oscillograph is the best instrument to record the EEG. In clinical examinations it is of value to use many pick-up points on the head and to amplify and display the potentials between pairs of electrodes, but the elementary features can be shown with a single recording channel.

The features of the operation of the ink-writer will be demonstrated to you. Because of the small size of the EEG potentials, they may be easily obscured by voltages from other sources such as the supply mains, muscle action potentials in the subject and interference from nearby electrical apparatus. The amplifiers connected to the ink-writers have circuits which magnify potentials between the two input leads (out-of-phase signals) but attenuate voltages (with reference to earth) which occur equally in both input leads (in-phase signals).

It is important to secure a low-resistance connection between the pick-up electrodes and the scalp. Part the hair in the left occipital region and also over the occiput and degrease the skin with alcohol. Apply electrode jelly to the electrodes and secure firmly to the scalp by means of the head-harness. The inter-electrode resistance should not be greater than 5000 ohms. Apply an earth electrode to the mastoid region. Start the recording with the subject's eyes shut and the amplifier gain control at minimum. Gradually increase gain until a suitably sized trace is obtained. Use a pencil to mark on the recording paper your instructions to your partner during the subsequent observations.

1. Record for two or three minutes with the eyes shut and note the alpha rhythm. Does it disappear spontaneously? Measure its frequency later.

2. Show the effect of opening and closing the eyes.

3. Show the effect of mental arithmetic.

4. Show the effect of distracting the subject's attention.

5. See if you can detect any difference in the effects on the alpha rhythm of the subject when he recalls (a) auditory and (b) visual memory.

CHAPTER NINE

ALIMENTARY CANAL AND KIDNEY

9.1 Human Salivary Secretion

First determine the resting rate of your salivary secretion over a period of ten minutes. Swallow any saliva in your mouth, start the stop clock and from now onwards do not swallow any saliva. By cheek movements collect the saliva towards the lips and expel it all through the tapered glass tube into a 10 ml. graduated cylinder. Make a saliva collection every two minutes and read the level in the cylinder one minute after addition of the sample. Ignore any froth. Plot your results as a graph and find the average rate of secretion of mixed saliva in ml. per minute. Now follow the response to the stimuli a, b and c. In each case follow the same routine and measure the response until the resting level has been regained.

(a) Chemical stimulation of the tongue: place a small amount of sodium chloride on the surface of the tongue.

(b) Mechanical stimulation during chewing: gently chew the rubber cork for one minute. Take the cork out and put it in the beaker provided.

(c) Electrical stimulation of the mucosa of the tongue: use the square wave stimulator and electrodes (two strips of silver foil mounted side by side on a plastic rod) to deliver a series of shocks at repetition rate of 5 per second for ½ minute to the surface of the tongue. Set the strength control at zero and gradually increase until shocks are felt. Do not allow the electrodes to touch metal fillings in your teeth as this can be painful. Display your results in the form of a graph.

9.2 In the preceding experiment the saliva collected contained contributions from a number of glands and is referred to as mixed saliva. It is possible to fit a device over the opening of the duct of the parotid gland so that the secretion from this gland alone is collected. The flow of saliva is affected by anxiety, so many people are unsuitable subjects for a demonstration of this collection method before their fellow students. Identify in your partner the opening of the parotid duct which lies on the oral surface of the cheek opposite the second upper molar tooth. Also identify the opening of the submandibular gland. This lies on the summit of the sublingual papilla. Use a small electric torch to illuminate the inside of the mouth. To collect saliva from the parotid duct a small plastic capsule is held over the duct opening by means of a suction ring. Saliva collects in the central compartment of the capsule and is led away by a fine plastic tube. The necessary adhesion is produced in the outer ring by withdrawing air from it into a syringe through another fine plastic tube. Measure the output of the gland in drops per minute and show the response to stimuli as in preceding paragraph. Conditioned responses, such as salivation at the sight of a slice of lemon can be elicited in some subjects.

9.3 Gastric Secretion

In this test the secretory response of the human stomach to a simple meal of thin oatmeal gruel or dilute alcohol (ethanol) is measured by the withdrawal of a timed series of small samples of the gastric contents through a slender flexible tube, which has previously been swallowed by the subject. Prior to the test the subject should take no food for twelve hours apart from a cup of tea two hours before. The procedure of swallowing the tube and its withdrawal at the end of the test must be carried out under the supervision of a member of the staff.

The subject should first examine the tube, note the weighted end, the tube perforations and the marks in the form of lines which indicate the approximate position of the weighted end when the marks are at the mouth, one line, the cardia; two lines, the fundus; three lines, the pylorus; and four lines the duodenum.

The tube is rinsed in dilute glycerine for lubrication and held by the subject in his right hand a few inches from the weighted end, the rest of the tube is held in his left hand. He then inserts the weighted end into his nostril and edges it gently inwards. When it reaches the pharynx the subject drinks some water and simultaneously advances the tube, it will slide easily down the oesophagus into the stomach. When the three line mark is reached, secure the end of the tube to the face with tape to avoid the possibility of swallowing the tube entirely. Insert a 20 ml. syringe into the end and gently remove the stomach contents, the volume is normally 20-100 ml. but of course will be altered by the amount of water swallowed during intubation.

The subject now swallows the test meal which is either 500 mls. of warm oatmeal gruel or 50 ml. 7 per cent. ethanol in water. As far as practicable the subject should avoid swallowing his saliva from now on as it will tend to neutralize the gastric acid.

Samples (10 ml.) of the gastric contents are withdrawn every 15 minutes for 2½ hours or until no further material is obtained. The samples are kept in a numbered series of test tubes and examined for bile and mucus. Test each sample for free and total acidity. The free HCl is measured by titration with 0.1N NaOH to about pH4 using Topfers reagent as indicator: the colour changes from pink to orange. It may be necessary to filter or centrifuge the sample before titration. A second titration is then carried out, this time to pH 8·5 using phenolphthalein as indicator to measure 'combined' acid. The added titres for free HCl and combined acid comprise the total acid of the sample.

Measure the total chloride concentration in each sample. Pipette 2 ml. of filtered gastric contents into a porcelain basin, dilute with 10 ml. water. Add 5 ml. 0.1N silver nitrate solution and 3 ml. of indicator (iron alum in concentrated nitric acid). The mixture is titrated with 0.1N thiocyanate solution until there is a permanent red-brown colour.

Express the results in the form of a graph and comment on your stomach's performance.

9.4 Mammalian Intestinal Muscle *in Vitro*

The apparatus consists of a central vessel of 100 ml. capacity with an outlet at its lower end for draining off the fluid. The central vessel is enclosed in a water jacket which is kept at 37° C by means of an adjustable electric heater. A hollow glass tube fitted with a hook at its lower end is clamped to a pillar by a boss head so that it can be lowered into the central vessel. The pillar also carries a balanced lever with a frontal writing point giving a magnification of about three or four times.

Fill the central vessel with Tyrode's solution (Fig. 9.1) at 37° C and the outer vessel with water from the tap at 37° C. See that the temperature of the inner vessel is maintained at 37° C by varying the position of the heater. Raise the hooked glass tube out of the central vessel. Have ready a Petri dish with Tyrode's solution and a threaded needle.

The rabbit will be killed for you. All the following procedures must be carried out with care to avoid stretching the gut. Remove gently, portions of jejunum minus mesentery. Syringe gently with Tyrode's solution through the pieces to remove the intestinal contents and at once place in a shallow dish containing Tyrode's solution at room temperature. Cut off a piece of gut about 3 cm long and pass a threaded needle through all the coats near one end from the inside to the outside; make a short loop. Pass another thread through the other end, tie and leave about 15 cm of thread. In tying the threads do not close off the lumen of the gut. Pass the loop over the hook and lower the tube down into the warm Tyrode's solution, lower the lever to a suitable position and fasten the long thread to the lever by pressing it into a small lump of plasticine. Adjust the weight (a piece of plasticine) on the lever so that the gut is just stretched. See that the gut is completely immersed. Attach the air supply pipe (or better the cylinder of oxygen with 5 per cent. carbon dioxide) to the hollow tube and adjust the screw clamp till about three bubbles per second pass through the fluid. This aerates and stirs the fluid.

Record the movements on a slowly moving drum. It may be necessary to wait a few minutes for the movements to become fairly regular. Keep the temperature at 37° C and see that the aeration is maintained.

While the lever is recording add to the bath 1 ml. of 10^{-5} adrenaline solution by means of a 1-ml. pipette. Make a mark below the record at the time of adding the solution. The concentration in the bath will be 10^{-7}. When an effect is obtained drain the central vessel and immediately refill it with fresh warm solution. If the dose of adrenaline is insufficient to produce an effect make up a 10^{-4} solution and add small quantities of this to the bath. Not more than 1 ml. of a solution of a drug should be added to the bath at any time to avoid sudden changes of temperature.

When the gut is contracting rhythmically once more, add acetylcholine (A.Ch.) to give a concentration of 10^{-7}. If an effect is obtained, drain off the solution and replace once more.

These observations should be repeated to give a trace for each partner. Then add to the bath a dose of A.Ch. which produces a well-marked contraction. When this

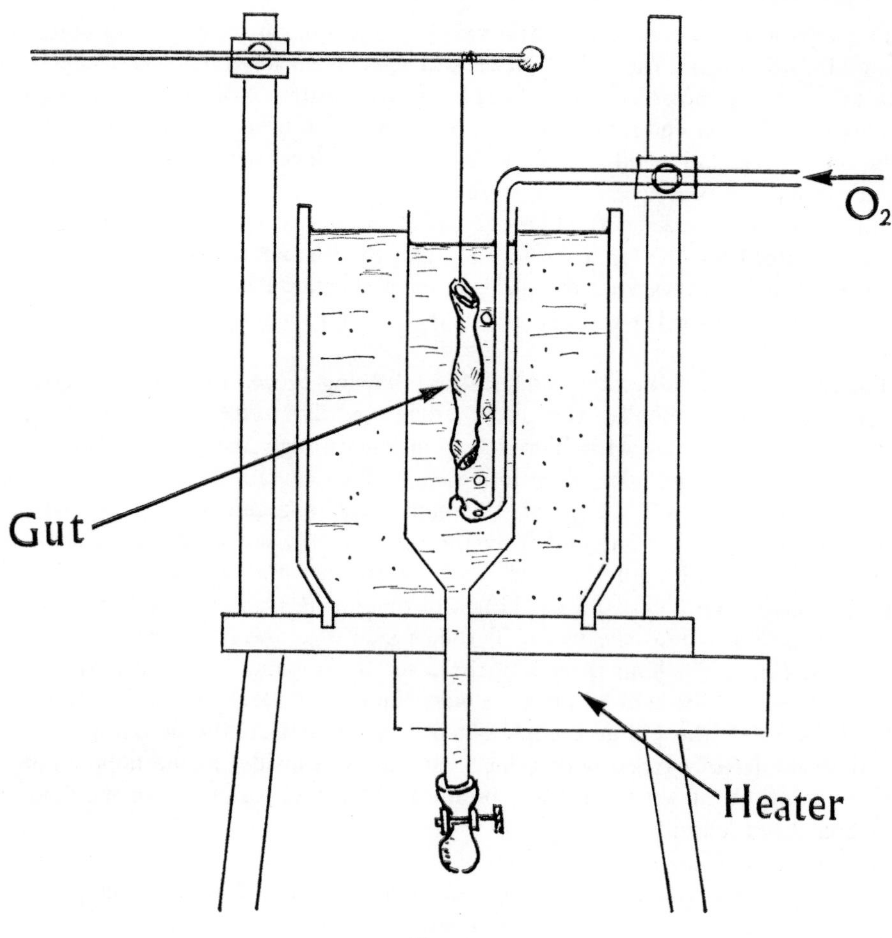

FIG. 9.1

has reached its peak add to the bath 1 ml. 10^{-4} atropine sulphate. This experiment will have to be repeated on a fresh piece of gut as the action of atropine is very prolonged.

Observe the effect of magnesium sulphate on the contractions.

What is the cause of the intestinal movements? How are they controlled? Discuss the reasons for keeping the temperature at 37° C and using Tyrode's solution.

9.5 The Control of Gut Activity by Mesenteric Nerves

A short piece (3 to 5 cm) of jejunum complete with mesenteric blood vessels is dissected out from a recently killed rabbit. The mesenteric nerves accompany the blood vessels but may be obscured from view by fat. A cotton thread is tied on the severed end of the main vessel, so that it can be guided onto the stimulating electrodes. At least 3 cm, preferably more, of mesenteric vessel should remain attached to the gut. The gut is set up in the organ bath as described in section 9.4 but in addition a stimulating electrode connected to the output of a square wave stimulator is immersed in the bath alongside the gut and the mesenteric attachment pulled through the electrode by the thread. Care should be taken that the mesenteric attachment is sufficiently slack to allow the gut movements to move the recording lever. Record the gut movements on the drum (speed 1 cm per min.). Set the stimulator thus: selector switch, PULSE OFF; strength, 10; pulse rate, 30 per sec; pulse length, long. Make a mark on the drum surface slightly ahead of the writing point and when the latter reaches it turn the selector switch to the REPEAT PULSE position, after one minute return switch to PULSE OFF position. The gut movements should be inhibited or reduced by the nerve stimulation. The effect should resemble the reponse to adrenaline added to the bath as seen in experiment 9.4. These fibres are referred to as adrenergic, read about this in your textbook.

Now make a series of stimulations, each of 1 minute duration with a rest period of two minutes between stimulations at the following pulse rates: 1, 5, 10, 20, 50, 100 per second. Does the pulse rate affect the magnitude of the inhibition of the gut movements?

9.6 Peristalsis in the Isolated Intestine

If guinea-pig ileum is set up in a bath of aerated Tyrode solution at 37° C as in paragraph 9.4 it will display very little spontaneous movement, but when the wall of the intestine is stretched slightly by a small positive pressure within the lumen it responds with peristaltic movements which travel caudally. The apparatus is arranged as in Figure 9.2. A piece of ileum 5 to 6 cm long is removed from a freshly killed guinea-pig and the caudal end tied over the glass tube connected to the pressure reservoir. Tyrode is run gently through the ileum from the reservoir to displace any air and the rostral end of the ileum is then closed with a ligature. A thread from the ligature transmits changes in length of the ileum to the frontal writing lever. The intralumenal pressure is determined by the relative levels of the Tyrode in the reservoir and the Tyrode bathing the preparation. Examine the rack-work which controls the height of the reservoir. Adjust the reservoir height to produce a positive pressure of 2 or 3 cm of water inside the ileum. A peristaltic wave will be seen to move down the ileum and expel liquid into the reservoir. This liquid movement is transmitted through an air column link to a piston recorder writing on the kymograph drum above the frontal lever registering the longitudinal movements.

Establish the threshold pressure for the peristaltic reflex. Note that the movements can be maintained for half a minute or more. Rest the preparation for two minutes at zero positive pressure between trials.

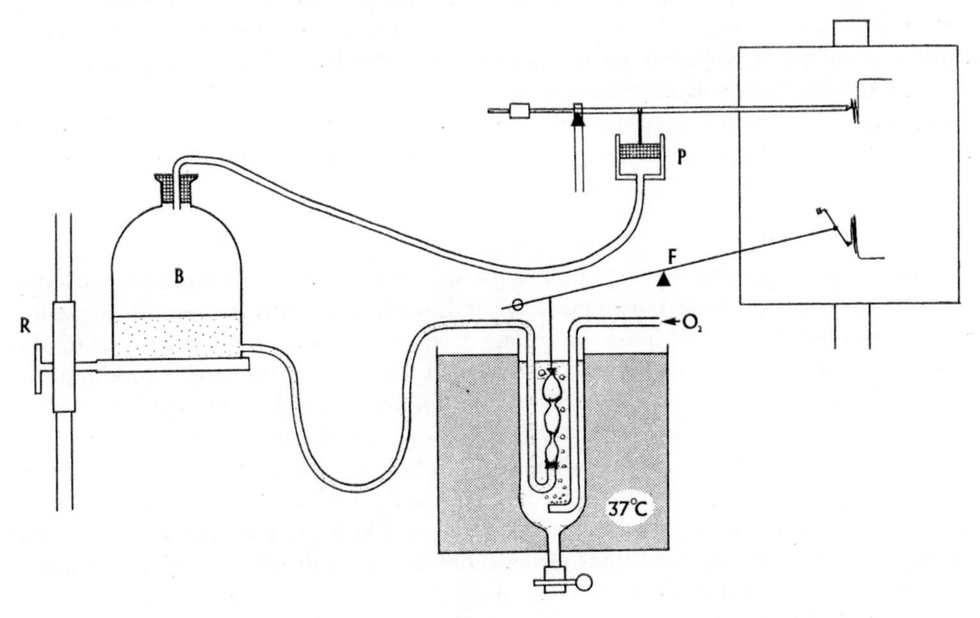

FIG. 9.2

Apparatus to show peristaltic reflex in isolated guinea-pig ileum. R, is a rackwork to vary the height of reservoir B and so determine the pressure in the ileum. Movement of liquid between the ileum and the reservoir is detected by the piston recorder P. Longitudinal changes are recorded by the lever F.

Since the response involves a reflex arc it is readily influenced by substances which alter nervous conduction or humoral transmission at synapses. Try the effect of a local anaesthetic such as Lignocaine on the preparation, start by adding 0·1 mg. to the bath and increase the dosage as necessary. Can you abolish the reflex? Wash out the anaesthetic, wait for the reflex to return and then try the effect of a ganglion blocking substance, hexamethonium. Add 0·1 mg. to the bath at first, increase the dose if ineffective.

9.7 Water Balance

When water is drunk it is rapidly absorbed by the intestine. The resulting dilution of the blood is sensed by osmoreceptors located in the brain and commands are sent to reduce the output of antidiuretic hormone from the posterior pituitary gland. The kidneys respond to the lowered level of hormone by increasing the rate of production of urine. This urine has a low specific gravity. Within limits, the greater the blood dilution the more profuse the diuresis. If however, a saline solution isotonic with blood is drunk instead of water, dilution of the blood does not follow. In the following tests the subject should arrange to make each test under approximately equivalent conditions, i.e. at the same time of day and following a standardized meal. The meal should be a light one with one cupful of fluid. In hot conditions rather more should be drunk, enough to maintain a low rate of urine production. One hour after the meal empty the bladder, one hour later empty the bladder again.

Measure the volume of urine and its specific gravity. This establishes the rate of urine secretion prior to the test.

(1) Drink 750 ml. of warm water. Empty your bladder at 20 minute intervals. Measure the volume and specific gravity. Express the results on a graph. How long does it take to excrete the quantity that you drank, i.e. restore the water balance?

(2) Drink 750 ml. of warm 1 per cent. sodium chloride solution. Empty your bladder at 30 minute intervals. Measure the volume and specific gravity. Plot your results on the same graph as in test (1). Comment on the response.

(3) Produce a marked diuresis by drinking three or four cups of hot tea (500-700 ml.). Plot the response as in test (1). Repeat the test on another day under equivalent conditions but, after the diuresis has begun, take a quarter of an hour of vigorous exercise, e.g. a game of squash racquets, work on the bicycle ergometer or fast running. Does the diuresis follow the same time course as before? Speculate on the possible mechanisms that may be involved.

9.8 Urine Sodium and Potassium Levels during Water Balance Tests

Read section 5.30 on the use of the flame photometer for estimation of sodium and potassium. Make appropriate dilutions of urine collected during tests (1) and (2) in paragraph 9.7 and express your results in graphical form. Comment on the significance of the results.

URINE DILUTION AND CONCENTRATION TESTS

In the normal kidney, the filtrate in passing through the tubules becomes hypertonic by reabsorption of water from the distal segment. Impairment of this part of the nephron results in failure of the concentrating power of the kidney. The failing kidney excretes the low threshold substances (urea, uric acid etc.) in lower concentration than in normal urine. If fluids are withheld for some time from a normal person, he passes urine of S.G. 1·030 or over. The impaired kidney is unable to raise the S.G. much, if at all, above that of protein-free plasma.

The failing kidney is also unable to reduce the S.G. of the urine below that of the filtrate, which suggests that the ability of the tubule to reabsorb sodium and chloride from an isotonic tubular fluid is impaired. Thus when a large quantity of fluid is drunk the S.G. does not fall to the low level (1·001-1·003) as in health but remains around that of deproteinized plasma (1·010). The fact that the urine is isotonic with protein-free plasma suggests that the tubular cells have lost their property of selective permeability and that substances are returned to the blood in the peritubular capillaries largely by a process of diffusion. This is usually a later manifestation of tubular impairment than loss of concentrating power.

9.9 Dilution Test

The subject should have a meal with a moderate allowance of fluid about 7 p.m. on the evening before the test and should thereafter take no further food or fluid.

On the day of the test, empty the bladder completely at 10 a.m. and discard the urine specimen. Collect the first test specimen at 11 a.m. and during the following 30 minutes drink 1200 ml. of water. Collect separate urine specimens at half-hourly intervals during the following 3 hours. It is important that the bladder should be emptied as completely as possible when each specimen is collected. Measure the volume and S.G. of each specimen. Make a graph showing the rate of fluid excretion over the test period.

Interpretation: about 1 litre of fluid should be excreted within the 3 hour period, the greater part during the first hour. The S.G. of the largest specimen will approach 1·002. When renal function is impaired the elimination will be reduced and the minimum S.G. will be comparatively high, e.g. 1·010.

9.10 Concentration Test

On the evening before the test the subject should have a meal before 7 p.m. with a high protein content but not more than 200 ml. of fluid; thereafter he should have no food or fluid until after the test is completed. Any urine voided during the evening or night should be discarded.

In the morning collect 3 separate urine specimens at intervals of one hour. Determine the S.G. of each specimen.

Interpretation: normally the S.G. of at least one of the specimens will exceed 1·025, figures over 1·030 being frequently obtained. Renal insufficiency is indicated by a S.G. consistently below 1·020 under the above conditions.

CHAPTER TEN

THE EFFECTS OF EXERCISE

The respiratory changes which accompany exercise have been considered in Chapters 3 and 4. This chapter is concerned with changes in the cardiovascular system and the skeletal musculature. An estimation of the metabolic cost of work is included. No person who has cardiac disease or who has had rheumatic fever should act as subject in these experiments.

Effects on Pulse and Blood Pressure

10.1 The subject sits quietly and his pulse rate is counted, by palpation of the radial artery, for 30 seconds every minute until three consecutive readings indicate that a stable base-level has been reached. The subject then performs strenuous exercise, e.g. stair-climbing at speed, until his respiratory embarrassment is marked. He now returns to the laboratory and resumes his seat. The time spent exercising is noted; pulse-counting is resumed as before and continued until the base-level has been regained and maintained for three minutes.

Prepare a graph showing the changes in Pulse Rate with Time with a gap for the period of exercise.

10.2 Repeat experiment 10.1 but measure the systolic and diastolic blood pressures by auscultation (or, less satisfactorily, the systolic pressure by palpation) instead of the pulse rate.

Graph these results showing systolic and diastolic pressure against time with a gap for the period of exercise. Shade in the pulse pressure area.

10.3 Changes in pulse rate and blood pressure during exercise are not easily monitored but the simple techniques of 10.1 and 10.2 can be employed during light to moderate exercise on a quiet-running egometer (see 10.9). This experiment is best performed with a group of four consisting of subject, separate recorders of pulse and blood pressure and a fourth member who supervises the ergometer.

The subject determines the maximum rate at which he can exercise without moving his upper limbs since the arm used for blood pressure studies must be still during estimations. He then rests quietly and basal values are established as before. The subject begins exercise and continues until three consecutive, similar readings of both pulse and blood pressure indicate that he has reached a new steady state. He then stops exercising and observations are continued until his pulse and blood pressure return to the base-line values. The changes in the first two minutes are particularly rapid (see 10.4); during this period the recorder of pulse rate should attempt counts for 15 seconds every 30 seconds. To ensure correlation of pulse and blood pressure readings, each recorder has a stop watch and these are started simultaneously.

Graph pulse rate and blood pressure against time as in 10.1 and 10.2.

10.4 The foregoing simple tests provide a general illustration of the way in which the cardiovascular system responds to exercise but the pulse rate changes can be followed in greater detail with an electrocardiograph. Recordings during exercise are rather unsatisfactory because of electrical interference from potentials arising in exercising muscles. This difficulty is avoided in this experiment by considering the heart rate from the moment that the exercise stops.

The subject sits on the bicycle ergometer and is connected to the electrocardiograph. He then rests quietly to establish a basal pulse rate. Record 20 sec at paper speed of 1 cm per sec. The electrocardiograph control switch is now held at the OBSERVE position with the signal mute ON to protect the stylus. Set paper-drive speed at 5 cm per sec.

The maximal pulse rates which can be attained during exercise are of the order of 180-200 beats per minute. To approach this level the subject must be given a heavy work task which he carries to exhaustion. When the subject has rested and his basal rate has been established, he is instructed to pedal at a fast rate against a heavy load for as long as possible. When he reaches 'breaking-point' he stops abruptly and thereafter sits as quietly as possible. The mute is switched off, the control switch moved to RECORD and continuous recording begun. After 30 seconds, reduce the paper-drive speed to 2·5 cm per sec, and after two minutes to 1 cm per sec. Continue recording for a total of six minutes or until the base level is regained. Switch off the electrocardiograph and examine the record.

The distance between successive beats is the reciprocal of rate. Measure the distances between beats at the beginning of your record; these should correspond to a rate of 150 per min or higher. If not, your subject has not made a maximum effort.

If the record is satisfactory, measure each interval for the first two minutes and prepare a graph with beat interval on the ordinate and successive beats along the abscissa. The rapid fall in rate is achieved by increasing vagal tone; in many subjects the rate does not come down in a smooth curve but exhibits a 'hunting' effect. Is this demonstrated in your subject? The experiment can be repeated to determine whether the oscillations are related to sinus arrhythmia. Once recording begins, observe the subject closely and use the marker to signal his respirations above the electrocardiogram.

Finally, from the record, work out a pulse-rate for each 15-second interval from the cessation of exercise and plot a graph of pulse rate against time.

10.5 Cardiovascular changes are not entirely a consequence of exercise, they may occur before the exercise begins. Fear or anxiety before an athletic contest have potent effects on the sympathetic nervous system.

The subject is connected to the electrocardiograph and is instructed to commence a Step Test (see 10.6) after a stated time interval without further command. He is provided with a stop-watch to allow him to do this. The electrocardiograph is switched to RECORD two minutes before this period is up and a continuous record taken till exercise begins, at a speed of 1 cm per sec. Then the instrument is switched off and a beat by beat analysis of the record made.

The pattern obtained will vary from subject to subject. Did your subject change his heart rate before the exercise began? Write a short note on the possible mechanisms underlying these effects.

10.6 Physical Fitness Tests are used as a basis for comparison between individuals and in the same individual at different times. As the cardiovascular system is a limiting factor in sustained exercise, its response to a short period of standardized exercise procedure is commonly used to assess fitness. These tests are sometimes referred to as Cardiac Tolerance Tests.

A frequently used procedure is the Step Test. This involves pulse counting and a step of suitable height. The description given is based on a Harvard Fatigue Laboratory test.

The subject steps on and off a platform 30 times per minute, in time to a metronome, with four distinct foot movements, namely (1) right foot on, (2) both feet on, (3) right foot off and (4) left foot off. He may lead with his left foot if he wishes and must straighten his back and knees between (2) and (3) and between (4) and (1). The step is 50 cm (20 in.) high for male subjects of average build; short-legged men, women, adolescents and the elderly are tested on a platform 42 cm (17 in.) in height. The test can be made more strenuous by making the subject carry weights.

Start a stop-watch at the beginning of exercise and time the duration of the subject's effort to the nearest second. If he consistently fails to keep time over 20 seconds he is stopped; he may stop voluntarily at any time and the observer ends the exercise after five minutes in every case.

The subject sits down immediately he ceases to exercise. After exactly one minute, recorded by the stop-watch, his pulse rate is counted for 30 seconds.

The duration of effort and pulse count are then applied to Figure 10.1 to give an index of fitness. Collect results from as many members of the class as possible and construct a histogram of the physical fitness of the group. Does the test confirm your pre-experiment assessment of each subject's physical fitness?

10.7 Local Effects of Prolonged Exercise in Muscle

It is a familiar experience that intense muscular work produces cramp-like pain in the muscles which limits the exertion. The pain is quite unlike the fatigue which brings less strenuous exercise to an end. It is due to accumulation of pain-producing

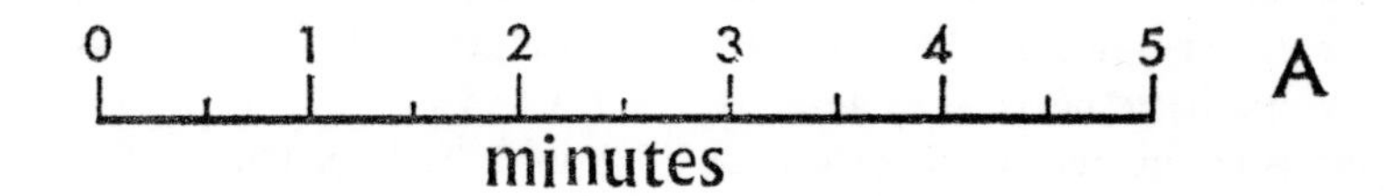

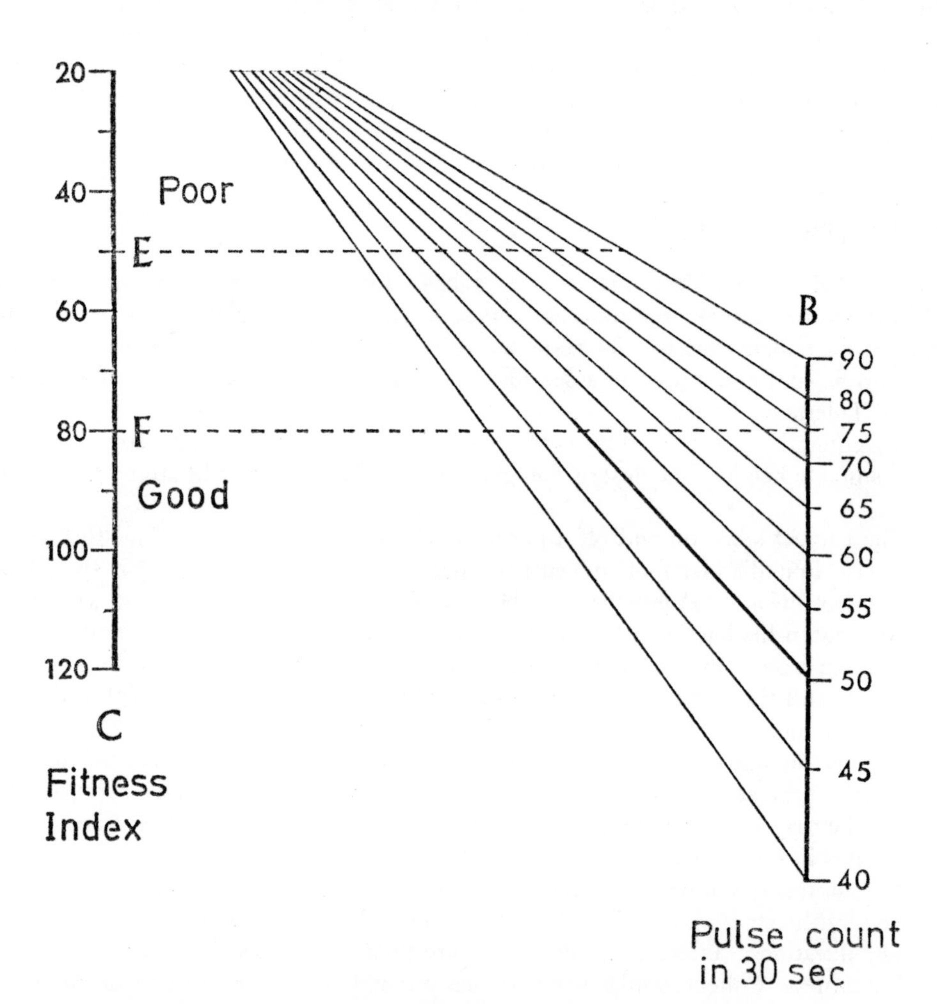

Fig. 10.1

(1) Mark duration of effort on line A. (2) Place a perspex setsquare with one edge along line A and its right angle at the point marked. (3) Determine the point of intersection of the vertical edge of the setsquare with the appropriate Pulse Count (line B). (4) Draw a horizontal line through this point intersecting line C. (5) This point is the index of fitness. The lines F and E indicate upper and lower limits for most normal subjects. The same table is used whether the subject has been tested on a 50 cm or a 42 cm step.

substances, metabolites, in the muscle which form during muscular contraction. A certain threshold concentration must be reached before pain occurs. This happens only when their rate of production exceeds the rate at which the circulation through the muscle can remove them. Hence pain, normally, only occurs with very intense muscular work. In more moderate work, even prolonged work, the circulation removes the substances as fast as they are formed, they do not accumulate and no pain occurs. If, however, the circulation is arrested before exercise begins the threshold concentration is built up quite quickly, even by mild exercise of the ischaemic muscles, and pain is produced.

1. Wrap a sphygmomanometer cuff round the left upper arm of the subject. Inflate the cuff to 180 to 200 mm Hg. The subject now alternately grips and relaxes a wooden rod with both hands once per second in time to a metronome. Record the time in seconds from the start of the exercise until the subject first experiences pain (it must be real pain as distinct from minor discomfort due to the cuff) and until he has to give up. Release the cuff pressure.

2. Now inflate a cuff on the right arm and repeat the experiment but at half the rate of working, i.e. one grip every two seconds. Record times as before. This time to produce pain is about twice as long as in the previous experiment. This shows that, with the circulation arrested, the total amount of work necessary to produce pain is constant and speed of working is not important. Release this cuff.

3. With the cuff deflated, the subject now grips at the fast rate (once per second) with his left hand for one and a half minutes. As soon as he stops work the cuff is inflated. Record the time from occlusion of the circulation until pain appears.

As in the previous experiments the pain persists as long as the circulation is arrested and disappears as soon as the cuff is released.

Arterial disease frequently results in partial blockage of the vessels with reduced blood flow. Then, quite mild exercise, e.g. a short walk, produces muscle pain which disappears with rest (intermittent claudication). The amount of work required, or distance walked, to elicit pain is remarkably constant in any given case. 'Angina pectoris' is the term applied to development of the ischaemic pain in cardiac muscle when the coronary arteries are partially occluded.

10.8(a) Mosso's Ergograph

Insert the first and third fingers into the fixed tubes and the second fingers into the ring. Adjust the subject's position till his forearm rests comfortably on the platform; strap the forearm in position. The ring is attached so that when the finger is flexed it pulls a writing lever over a drum and at the same time raises a 3 kg. weight (a smaller weight, say 2 kg., should be used with female subjects). Lay the drum on its side, screen it from the subject and select speed 1·5 mm per sec. Provide a signal every two seconds, e.g. a flashing lamp. The subject flexes his finger at this rate until fatigue is so great that the weight can no longer be moved.

Describe the shape and duration of the record obtained and draw conclusions from it.

Now demonstrate the value of rest pauses. Repeat the experiment (either use the other hand, or the same hand after an hour's interval) but allow the subject to rest for five seconds every minute. This is equivalent to a rest period of five minutes every hour which is the interval adopted by the army.

Measure the total lengths of all the lines in each experiment and calculate the total work done.

$$\begin{aligned} \text{Work} &= \text{force} \times \text{distance} \\ \text{force} &= \text{weight in kilograms} \\ \text{distance} &= \text{sum of weight lines} \end{aligned}$$

Express answer in kg. metres.

10.8(b) The Finger Dynamometer

In this apparatus the weight of Mosso's ergograph has been replaced by a graduated spring balance. The force opposing the finger movement therefore increases linearly with finger displacements. Tell the subject to make a series of maximal contractions in time with the signal. The length of the lines on the kymograph trace will record the maximum tension developed. It will decrease as fatigue occurs. Calibrate your record with a force scale, taken from the graduations on the spring balance. Show the effect of rest pauses as in 10.8(a). Express your results in terms of decline of maximum contraction force.

10.9 The Metabolic Cost of Work

The most convenient apparatus for measuring the work done in exercise is the bicycle ergometer since it brings a large mass of muscle tissue into action. The oxygen consumption of the subject can be measured either by collecting his expired air over a timed period and measuring its oxygen content and its volume or by connecting the subject to an oxygen-filled Benedict-Roth spirometer (Fig. 4.3).

Two types of bicycle ergometer are in common use. In one the subject rotates a fly-wheel against a band-brake and in the other he drives a dynamo, the output of which is measured electrically. The procedure of the experiment is the same for both types of ergometer; the only difference is the method of calculation of work output.

In this experiment we assume that the additional oxygen consumed during exercise, that is the oxygen consumption during exercise minus the resting oxygen consumption, is all used by the muscles engaged in pedalling. Since the calorific value of oxygen is known, the energy represented by this oxygen increment can be calculated. Only a fraction of this energy does external work; much of it appears as heat in the exercising muscles and is dissipated at the body surface. The efficiency of exercise is expressed as the amount of energy emerging as external work relative to the amount of extra energy expended.

Thus if E_R = energy used at rest in 1 min, E_W = energy used at work in 1 min, and W = work done in 1 min

$$\text{Net efficiency} = \frac{W}{E_W - E_R}$$

$$= \frac{\text{Work done during exercise}}{\text{Extra energy expenditure during exercise}}$$

This fraction is always less than unity and is usually expressed as a percentage.

PROCEDURE

A group of four is convenient for this experiment. The efficiency of muscular exercise depends, amongst other things, on the rate of doing work but since individuals vary in their physique, no fixed rate can be recommended. The subject should pedal in time with a metronome set at 100 beats per minute (1 beat per leg movement) and adjust the load to obtain a level of exercise which he considers he could maintain without distress for fifteen minutes. It is important not to set the rate of working at too high a value, otherwise the subject builds up an oxygen debt (see 10.10). This produces falsely high values for efficiency and an apparent efficiency above 20 per cent. can almost always be traced to this cause.

The subject now rests for at least 20 minutes to allow his oxygen consumption to return to the resting level. Meanwhile, the other members of the group carry out trials to find their own suitable work loads.

When the subject has rested, he sits quietly on the bicycle while his oxygen consumption is measured, either by collection of his expired air in a Douglas bag over a timed period, cr by means of a Benedict-Roth spirometer. He then cycles steadily, in time to the metronome or keeping the speedometer needle steady at the indicated point. After five minutes he is assumed to be in equilibrium at the exercise level and his oxygen consumption is recorded over the next five minutes.

CALCULATION OF WORK DONE

Band-brake Ergometer. The work done is the product of the force acting on the wheel and the distance moved by a point on the rim. The force against which the rim of the wheel moves is shown directly on the 'Monark' bicycle[29] by a pointer on a scale 0–7kp (1 kp equals 1 kilogram weight). On other types of bicycle the force is the difference in tension shown on spring balances at either end of the band-brake. The force must be measured when the subject is working, and, if necessary, adjustments should be made to keep the force constant during the exercise period. The distance moved by a point on the rim of the wheel is the product of the circumference of the wheel and the number of revolutions. In the case of the Monark bicycle, one complete turn of the pedal wheel moves the brake wheel 6 metres, so if the subject pedals to the time of a metronome set at 100 beats per minute, so that each leg movement (i.e. half a pedal revolution) synchronizes with a metronome beat, the pedal wheel revolves at 50 r.p.m. and the brake wheel travels $50 \times 6 = 300$ metres per minute. Thus if the tension is set at 1 kp, this means the subject is working at a rate of 300×1 kg metres per minute.

Work done per minute = Force × Distance per minute. Examine your bicycle to see how force and distance are measured.

Dynamo Ergometer. The output of the dynamo ergometer is measured with a voltmeter and ammeter, the product of the meter readings, volts × amperes, giving the output in watts. Allowance must be made for mechanical and electrical losses in the gears and dynamo. It has been found experimentally that the apparatus is usually about 70 per cent. efficient. This has nothing to do with muscular efficiency; it means that the output in watts calculated from the meter readings is only 70 per cent. of the power applied to the pedals which is the figure we require. Since 1 kcal per minute is equivalent to 70 watts and also to 427 kg. metres per minute, the work done is readily converted to kg. metres. The watt is a measure of power, however, and the machine really gives the output at an instant of time. Since oxygen consumption is measured over several minutes, it is important to maintain the output as constant as possible during the sampling period.

CALCULATION OF OXYGEN CONSUMPTION

Douglas Bag Method. Collect in a Douglas bag the expired air from the subject during a timed period. Withdraw a gas sample from the bag into a paramagnetic oxygen analyser[18,23] and estimate the oxygen content as described in section 3.25. A Haldane gas analyser may be used instead but this requires some skill. Use the nomogram in Figure 4.1 to find the calorie value of a litre of expired air at the oxygen content you have found in your sample. Expel the expired air from the Douglas bag through a gas meter and divide the volume by the collection time to give the volume of air expired per minute. Convert this volume to STP with the nomogram in Figure 3.2. Multiply the expired air volume per minute by its calorific value per litre and this gives the energy expenditure. Find this value for the subject at rest (E_R) and while working (E_W). Convert the energy values from k calorie to kg. metres: 1 k calorie equals 427 kg. metres and substitute in the equation on p. 254 to obtain net efficiency.

Benedict-Roth Method. This apparatus is described in section 4.2. Oxygen consumption per minute is obtained from the spirometer chart and converted to STP using Figure 3.2. Convert from oxygen consumption per minute to k calories per minute: 1 litre of O_2 is approximately equivalent to 5 k calories. Then convert to kg. metres: 1 k calorie equals 427 kg. metres. Substitute in equation on p. 254 to obtain net efficiency.

It is interesting to compare the result with measurement of efficiency in engines; a good steam engine is about 10 per cent. efficient, the best petrol engine about 30 per cent. efficient. What was your horsepower while cycling? One horse power is 0·178 k calorie per second = 4560 kg. metres per minute = 746 watts.

Repeat the experiment on each member of the group. Compare the results and write a short note on your findings.

10.10 Oxygen Debt

A sprinter performs a large amount of work when he runs 100 yards but the increment in his oxygen uptake, while he is actually running, is very small. If calculations of work done and energy consumed were based on these figures an efficiency in excess of 100 per cent. might even be attained, an obviously bizarre result. However, the sprinter shows considerable respiratory distress after his race; he must 'pay the metabolic piper' and his oxygen intake is enhanced for many minutes. He is considered to have contracted an Oxygen Debt and the extent of this deficit is the real measure of the energy cost of his sprint.

In metabolic terms the energy output of the muscles exceeds the capacity of the aerobic mechanisms and anaerobic pathways make up the difference; later, additional oxygen is required to reconstitute these anaerobic sources of energy at their pre-exercise level. At low levels of activity oxygen demand does not exceed supply and no oxygen debt is incurred; hence the reason for stressing a low work-rate in 10.9.

Between these extremes there is a steady gradient of increasing oxygen debt.

The experiment is performed as in 10.9 but the subject is given strenuous work to perform.

The subject's resting oxygen consumption is established. At a predetermined signal, he begins moderate to severe work and recording of his oxygen uptake begins simultaneously. He stops exercising after one minute (or 30 seconds with very severe work) and sits quietly on the ergometer. He remains connected to the spirometer, however, for 15 min. or until his oxygen consumption returns to its resting level. This may necessitate refilling the spirometer during the experiment but if this is performed quickly and the spirometer is not halted during the procedure, no significant error will be introduced.

The work performed is derived as in 10.9.

Examine the spirometer record and divide it into 30 second intervals. Calculate the total oxygen consumption in each period and subtract the total expected (resting) oxygen consumption from each. The sum of these differences is the total oxygen increment for exercise. Convert it to a rate of litres per minute of work and calculate the nett efficiency in the usual way.

Now calculate an efficiency based solely on the oxygen consumption during the period of exercise and demonstrate the gross error inherent in this manoeuvre.

10.11 Temperature Regulation during Exercise

When a subject carries out hard physical work, his oxygen consumption is greatly increased, e.g. he may consume 0·3 l. O_2 per min at rest and 2 l. per min when working hard. Nearly all the energy liberated appears as heat as can be shown in paragraph 10.9 where even under favourable conditions the efficiency of muscular exercise is found to be between 10 and 20 per cent. In terms of heat liberation, this working level roughly corresponds to the output of a 600 watt electric heater. The temperature regulating centre in the central nervous system will limit the temperature rise by increasing the

rate of heat loss through the skin. Blood is diverted to cutaneous vessels, the skin temperature rises, and sweating may occur. The amount of heat stress will be affected by the severity of the work, the amount of clothing worn and the cooling power of the environment (see Chap. 11). In the experiment to be described the subject performs moderate work on a bicycle ergometer and the cardiovascular and sudomotor responses are measured.

Measurements

Skin Temperature. Draw circles on (a) the forehead (b) the pulp of the middle finger. Use a thermistor or thermeolectric thermometer (see 11·5) to measure temperature at these points.

Hand Blood Flow. Use either venous occlusion plethysmography (6.44) or calorimetry (6.54) to measure hand blood flow. It may be necessary to stop cycling during plethysmographic estimations.

Body Temperature. Use a clinical thermometer as described in 4.5. The subject should keep his mouth closed and breathe through his nose. If this proves impossible because of the increased ventilation during exercise mouth temperatures will be depressed below core body temperature.

Heart Rate. Count the pulse at the wrist or use the electrocardiograph. There will be interference between electromyographic and electrocardiograph potentials and chest leads, so pauses for recording may be necessary.

Sweating. Note time of onset and cessation.

Procedure

Take measurements while the subject, clad in his normal clothing, sits at rest on the bicycle. When base-line values have been obtained, give the signal to start cycling. Adjust the loading of the ergometer to suit the physique of the subject. Take measurements every few minutes until he begins to sweat and is red in the face. Stop the exercise and record the return to base line values.

Presentation of Results

Produce a graph in which the various quantities are plotted out in different colours with time as the horizontal axis. Calculate the rate of working and the work done during the exercise period from the dials of the ergometer; the method is described in paragraph 10.9. Estimate the heat production rate using the value of efficiency obtained in paragraph 10.9 or if this is not available assume efficiency of 15 per cent.

CHAPTER ELEVEN

ENVIRONMENTAL PHYSIOLOGY

The Cooling Power of the Environment

In man the body temperature is regulated chiefly by variation of heat loss. Loss of heat from the body occurs by (1) conduction and convection, if the air temperature is lower than that of the body; (2) radiation, if the temperature of the environment is lower than the exposed skin surfaces; (3) evaporation of sweat, if the air is not completely saturated with moisture. Since the mechanism of heat loss is so complex, it has not been possible to find a simple device for the estimation of the cooling power of the environment on the human body.

11.1 Atmospheric Humidity

The evaporation of sweat is dependent on the degree of saturation of the surrounding air with water vapour and on the air movement. The degree of saturation is usually given as the relative humidity, that is the actual water-vapour pressure in the air expressed as a percentage of the water-vapour pressure in air saturated with moisture at the same temperature. It is estimated by means of some form of wet and dry bulb thermometer. When the bulb of an ordinary thermometer is covered with a wick soaked in water, the water evaporates, unless the air is already saturated (relative humidity 100 per cent.), and lowers the temperature of the wet bulb until equilibrium is reached. The simple wet and dry bulb thermometer can give results with sufficient accuracy for many physiological purposes (see Nomogram Fig. 11.1).

To obtain reliable results it is necessary to have a rapid current of air over the wet bulb. This is obtained in the whirling psychrometer by whirling the apparatus and in the Assmann psychrometer by means of the fan. Examine the whirling psychrometer.

Study the scales so as to facilitate subsequent reading. Soak the wick and fill the reservoir with distilled water. Whirl the psychrometer rapidly for half a minute. Stop whirling, and quickly read the wet-bulb thermometer and then the dry-bulb thermometer. Note the readings and repeat until these are steady. Record the steady wet-bulb and dry-bulb temperatures. Using these read off the relative humidity from the accompanying Nomogram (Fig. 11.2). Empty the reservoir.

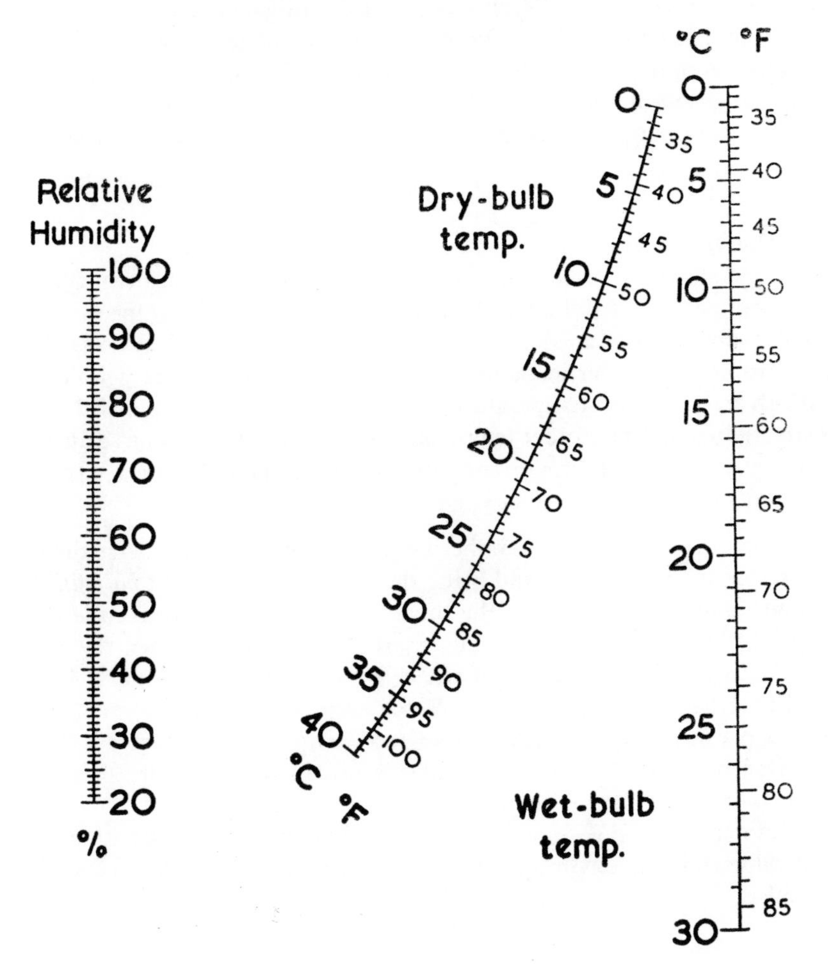

FIG. 11.1

Nomogram for finding the relative humidity of the air under calm conditions. This nomogram is applicable to the ordinary wet and dry bulb thermometer—not the whirling hygrometer.

(From Bell, G. H. and Weir, J. B. de V. (1947). *Brit. Med. J.*, ii, 174).

Examine the Assmann Psychrometer. Carefully remove the radiation shield. Thoroughly moisten the wick on the wet-bulb thermometer by raising a test-tube of distilled water round it. Replace the radiation shield. Start the motor. Read the two thermometers at one-minute intervals until the readings are steady. Record the wet-bulb and dry-bulb temperatures and find the relative humidity from the nomogram in Figure 11.2.

11.2 Katathermometers

These are alcohol thermometers with large bulbs. There are two types. The first has a polished glass bulb and contains alcohol coloured red. This measures the cooling power of the environment at body temperature and gives an indication of the suitability of the environment for particular kinds of work. The other has a bulb coated with polished silver, the high-temperature, silvered katathermometer. Silvered thermometers are graduated for different temperatures, one with blue-coloured alcohol between 125° and 130° F and the other with magenta-coloured alcohol between 145° and 150° F; they are used to measure air velocity.

(a) Hang the polished glass katathermometer on its stand, immerse the lower end in a mug of water at about 120° F and bring the alcohol up so that it *half-fills* the dilation at the upper end; quickly remove the mug. Overheating will break the thermometer. Dry the bulb. Do not allow it to swing. Start the stop-watch when the alcohol passes the 100° F mark and stop the watch when it falls to the 95° F mark. Repeat several times and take the average. The Kata factor engraved on the thermometer gives the millicalories per sq. cm lost in falling from 100° to 95° F. If the factor is divided by the number of seconds the 'dry Kata' reading is obtained, a measure of the cooling by conduction, convection and radiation.

The human body does not behave by any means like a katathermometer but empirical values relating environment to suitability for various kinds of work have been recommended.

	Dry Kata
Sedentary Work	6
Light muscular work	8
Heavy muscular work	10

(b) Use the thermometer (silvered) as in the preceding paragraph. Calculate the air velocity using the nomogram supplied for the thermometer.*

* Medical Research Council (1946). *M.R.C. (War) Memor.* No. 17, Suppl.

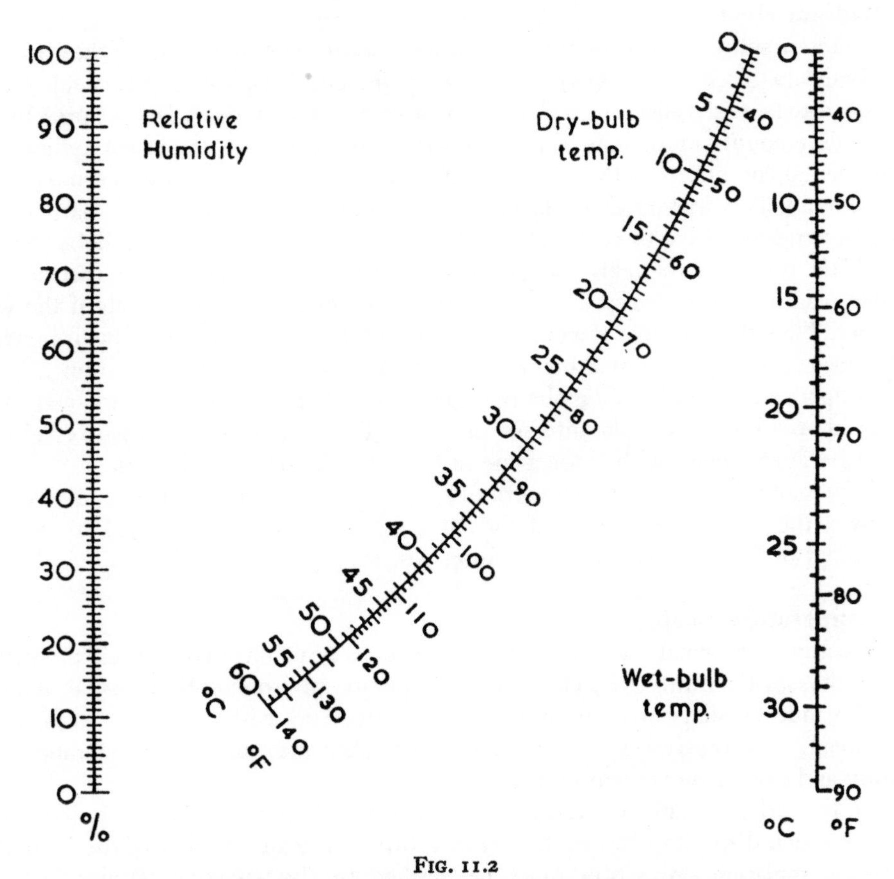

FIG. 11.2

Nomogram for finding the relative humidity of the air from the dry-bulb and wet-bulb readings of a ventilated psychrometer.

(Weir, J. B. de V. (1949). *Brit. Med. J.*, i, 257).

11.3 Radiant Heat

A human subject absorbs and also radiates heat. His skin and clothes act as a 'black body' (as used in Physics). An ordinary mercury-in-glass thermometer on the other hand acts as a reflector, and is insensitive to radiant heat. In the presence of appreciable amounts of radiant heat such a thermometer is an imperfect indicator of environmental conditions as they affect human beings. If, however, the thermometer is enclosed in a hollow metal sphere, usually 15 cm in diameter, with a black matt surface, then it becomes sensitive to radiant heat. Such an instrument is called a globe thermometer. If the walls of a room are warmer than the air of the room the globe thermometer will give readings above those of an ordinary thermometer and conversely if the walls are colder. Thus the readings of a globe thermometer are better indices of environmental conditions as they affect human beings than are ordinary thermometer readings.

Set up a stand with both a globe thermometer and an ordinary thermometer. Take readings at 10 minute intervals until successive readings are steady. Compare conditions on the laboratory bench with those 1 metre from a radiator or in the sun.

If the globe thermometer temperature, the air temperature and the air velocity are known, the mean temperature of the surroundings can be ascertained from charts 4, 5, 6 and 7 of M.R.C. War Memorandum, No. 17.

11.4 Temperature Scales

To compare thermal environments, it is necessary to have some scale of warmth which expresses the summated effect of the separate environmental factor in a single index. An index which with one limitation gives good measure of the warmth of the environment, is **Effective Temperature**. This takes into account the temperature, humidity and rate of movement of air.*

The limitation of effective temperature is that it does not allow for radiant heat. If there is much disparity between the temperature of the air and that of the surroundings, some radiation correction must be applied to the effective temperature. An approximate correction can be made by using the globe thermometer temperature instead of the air temperature to calculate the effective temperature. The figure so obtained is the **Corrected Effective Temperature.**

There are two scales of effective temperature: one, the basic scale, relates to persons who are stripped to the waist; while the other applies to people wearing light indoor clothing. This latter scale should be used.

Appraisal of Results. It is recommended that, wherever practicable, the corrected effective temperature in spaces where men live and work should be kept below 80° F.

* Medical Research Council (1946). *M.R.C. (War) Memor.* No. **17**, Suppl.

11.5 The Response to Body Warming

The temperature of the core of the body is regulated by the central nervous system. In temperate conditions the regulation is largely applied to the rate at which heat is lost from the skin. Skin heat loss, by radiation, convection, conduction and sweating is controlled by the blood flow through the cutaneous blood vessels and messages in the sudomotor nerves. If the body is attempting to excrete heat maximally, i.e. if there is 'heat stress', the great increase in skin blood flow can only be achieved by an increased pumping action of the heart, so that even though the subject is at rest physically, his heart rate and cardiac output will be raised.

In this experiment the activity of the temperature regulating centre is disclosed by abruptly changing the environmental conditions of the subject, so that if core body temperature is to remain constant, heat excretion must be increased. The circulatory adjustments which are necessary to increase heat excretion are detected by measuring heart rate, blood flow through the hand and skin temperature. Core temperature is measured throughout and sweat production and skin colour observed. In the ideal experiment all these changes are recorded and correlated from a single subject during a period of warming. A control period of observation is of course necessary to establish baseline values. However this full experiment will involve six or more observers and in some practical courses it may be found more convenient to study one or two responses only.

Recording

Observers should synchronize their watches and affix a time to each measurement so that at the end of the experiment a graph with an accurate time-axis can be prepared.

Heart Rate. Apply electrodes to arms or chest, connect to an electrocardiograph and switch on ink-writer for 20 seconds every two minutes, alternatively count the pulse at the wrist every two minutes.

Skin Temperature. Draw an ink-circle on the forehead, back of the hand and pulp of finger. Measure the temperature with the electrical thermometer at these points every three minutes.

The device used to measure skin temperature must be of small dimensions so that it does not interfere with the flow of heat from the skin to the environment, and of low thermal capacity so that it will rapidly follow temperature changes. The ordinary mercury-in-glass thermometer is not suitable, but two electrical methods are available, the thermistor and the thermocouple. The thermistor is a minute semiconductor resistance with a large negative temperature coefficient; the temperature is obtained from a calibration curve relating current flow in μA through the temperature-sensitive resistance with temperature. (See Fig. 1.5).

The copper-constantan thermocouple generates an electromotive force which will produce a current depending on the difference in temperature between the 'hot' and 'cold' junctions. For a cold junction, it is sufficiently accurate to use a couple immersed in water in a vacuum flask at about room temperature. The thermoelectric current is measured with a galvanometer and calibrated against a mercury thermometer by immersing both couple and thermometer in a stirred water-bath.

Body Temperature. Measure oral temperature with a clinical thermometer as described in paragraph 4.5. Take readings every five minutes, shake down, and replace the thermometer in subject's mouth.

Blood Flow. Measure the blood flow through one arm by venous occlusion plethysmography as described in paragraph 6.44. Take measurements every five minutes.

Sweating. Inspect the skin of the subject's face and record time of onset of sweating.

Skin Colour. Inspect skin of face and hand and note colour changes.

PROCEDURE

Control Period. Arrange the environmental conditions so that the subject feels cool or cold, e.g. let him sit with his jacket off in the draught from a fan. Take observations over a period of 15 minutes or until a steady baseline is obtained.

Warming Period. The subject should immerse his feet and calves in a bin of warm water (44° C). The water must not be so hot as to cause pain otherwise vasoconstriction will occur. Arrange his chair so that he can sit comfortably with the water almost up to his knees. Hot water should be added as required to maintain the temperature at 44° C. Wrap blankets round the subject. Take observations until an obvious response occurs, including sweating. This warming phase may take half an hour or more. Question the subject about how he feels and record his answers.

Cooling Period. Remove the bin and blankets, direct a cool draught towards the subject. Make measurements until baseline levels are regained.

PRESENTATION OF RESULTS

The various quantities should be plotted as lines of different colours on a single graph with time as the horizontal axis.

11.6 The Assessment of the Environment in Conditions of Overcrowding

Man is able to control his immediate environment in temperate latitudes by use of clothing and this, together with cardiovascular control of heat excretion through the skin, permits thermal comfort within a range of environmental conditions. In conditions of severe overcrowding, the excretion of heat, water vapour and carbon dioxide alters the environment itself and the physiologist may be required to assess, in human terms, the efficacy of temperature, and ventilation control systems. A suitable introductory exercise is to examine conditions in an overcrowded and poorly-ventilated lecture theatre, if such can be found.

Measurements

Air Temperature. Mercury thermometers mounted at head and foot level at both the upper and lower parts of the theatre should be read every 15 minutes.

Humidity. Measure with whirling psychrometer every 15 minutes.

Carbon Dioxide Concentration. Take samples with evacuated gas sampling tubes every 15 minutes.

Radiant Heat. Use a globe thermometer.

Physiological Responses in the Subject. Make these measurements on a sample member of the audience. Use a portable thermistor thermometer to measure skin temperature on forehead and pulp of middle finger every five minutes. Inspect face for sweating. Measure heart rate every five minutes. Record subjective description of the environment every quarter hour.

Presentation of Results

Baseline values should be obtained just before the audience assembles and observations continued for the period of the lecture. The various quantities should be plotted as lines of different colours with time as the horizontal axis. Estimate the total heat production of the audience (an adult at rest excretes about 70 kilocalories per hour). What would be the equivalent power of an electrical heater in kilowatts (1 kilocalorie/sec =4·19 kW)? Write an assessment of the ventilation and heating systems on the basis of your observations.

APPENDIX A

Subscript Number	*Name and Address of Suppliers*
1	C. F. Palmer, Ltd., 16 Carlisle Road, London, N.W. 9
2	Gallenkamp & Co. Ltd., Christopher Street, London, E.C.2.
3	A. R. Horwell, Ltd., 2 Grangeway, London, N.W.6.
4	Telequipment, Ltd., 313 Chase Road, Southgate, London, N.14.
5	Tektronix, Ltd., Beaverton House, Harpenden, Herts.
6	S.T.C. Transistor Division, Footscray, Kent.
7	Amplivox, Ltd., Beresford Avenue, Wembley, Middlesex.
8	Watson & Sons, Ltd., Barnet, Herts.
9	British Drug Houses, Poole, Dorset.
10	W. R. Prior & Co. Ltd., Bishops Stortford, Herts.
11	Devices, Ltd., Welwyn Garden City, Herts.
12	Siebe, Gorman & Co., Ltd., Tolworth, Surbiton, Surrey.
13	Zentralwerkstatt, Göttingen, W. Germany.
14	Bausch & Lomb Ltd., Mill Hill, London, N.W.7.
15	Henley's Medical Supplies Ltd., Clarendon Road, Hornsey, London, N.8.
16	Evans Electroselenium Ltd., Halstead, Essex.
17	Macfarlane & Robson Ltd., Hedgefield House, Blaydon-on-Tyne, Co. Durham.
18	Beckman Instruments, Queensway, Glenrothes, Fife.
19	Mullard Ltd., Torrington Place, London, W.C.1.
20	Airmed, 16 Wigmore Street, London, W.1.
21	Aimer Products Ltd., 56 Rochester Place, London, N.W.1.
22	G. T. Gurr, New Kings Road, London, S.W.6.
23	Servomex Controls Ltd., Crowborough, Sussex.
24	William Patterson & Sons, 57 Spring Gardens, Aberdeen.
25	Cambridge Instrument Co. Ltd., 13 Grosvenor Place, London, S.W.1.
26	Godart Ltd., Maidstone, Kent.
27	Electrophysiological Instruments Ltd., Bryans Newtongrange, Midlothian.
28	Mercury Electronics Ltd., 640 Argyle Street, Glasgow, C.3.
29	Cykelfabriken Monark, Varberg, Sweden.
30	Avo Ltd., Avocet House, 92-96 Vauxhall Bridge Road, London, S.W.1.
31	Ether Ltd., Caxton Way, Stevenage, Herts.
32	Solartron Electronic Group, Farnborough, Hants.
33	Scientific & Research Instruments Ltd., 335 Whitehorse Road, Croydon, Surrey.
34	Isleworth Electronics, Frederick Street, Waddesdon, Bucks.
35	Airmec Ltd., High Wycombe, Bucks.
36	Parkinson Cowan Ltd., Talbot Road, Streford, Manchester.
37	Plysu Industries Ltd., Woburn Sands, Bletchley, Bucks.
38	Grass Instruments, Quincy, Mass. U.S.A.

A general enquiry service is run by:

The Scientific Instrument Manufacturers' Association of Great Britain Ltd.,

20 Peel Street, London, W.8.

APPENDIX B

Bibliography

General

Bures, J., Petran, M. & Zachar, J. (1967). *Electrophysiological Methods in Biological Research*, 3rd ed. London: Academic Press.

Burn, J. H. (1952). *Practical Pharmacology*. Oxford: Blackwell.

Catton, W. T. (1957). *Physical Methods in Physiology*. London: Pitman.

D'Amour, F. E. & Blood, F. R. (1948). *Manual for Laboratory Work in Mammalian Physiology*. Chicago: University Press.

Dickinson, C. J. (1950). *Electrophysiological Technique*. London: Electronic Engineering.

Donaldson, P. E. K. (1958). *Electronic Apparatus for Biological Research*. London: Butterworth.

Edinburgh University Pharmacology Department (1968). *Pharmacological Experiments on Isolated Preparations*. Edinburgh: Livingstone.

Gay, W. I. (1965). *Methods of Animal Experimentation*. London: Academic Press.

Geddes, L. A. & Baker, L.E. (1968). *Principles of Applied Biomedical Instrumentation*. London: J. Wiley & Sons Inc.

Greene, E. C. (1955). *Anatomy of the Rat*. New York: Hafner Publishing Co.

Hale, L. J. (1958). *Biological Laboratory Data*. London: Methuen.

Hill, D. W. (1965). *Principles of Electronics in Medical Research*. London: Butterworth.

Hunter, D. & Bomford, R. R. (1960). *Hutchison's Clinical Methods*. 13th ed. London: Cassell & Co.

Kay, R. H. (1964). *Experimental Biology; Measurement & Analysis*. London: Chapman & Hall.

Lewis, T. (1950). *Exercise in Human Physiology*. London: Macmillan.

Nastuk, W. L. (1962). *Physical Techniques in Biological Research*, Vol. 4. London: Academic Press.

Nastuk, W. L. (1964). *Physical Techniques in Biological Research*, Vol. 5, part A. London: Academic Press.

Riggs, D. S. (1963). *The Mathematical Approach to Physiological Problems*. Baltimore: Williams & Wilkins Co.

Stott, F. D. (1967). *Instruments in Clinical Medicine*. Oxford: Blackwell.

Sidowski, J. B. (1966). *Experimental Methods and Instrumentation in Psychology*. New York: McGraw-Hill Co.

Venables, P. H. & Martin, I. (1967). *A Manual of Psychophysiological Methods*. Amsterdam: North Holland Publishing Co.

Whitfield, I. C. (1960). *An Introduction to Electronics for Physiological Workers*, 2nd ed. London: Macmillan.

Chapter 2: Muscle and Nerve

Katz, B. (1966). *Nerve, Muscle & Synapse*. New York: McGraw-Hill Co.

Tasaki, I. (1953) *Nervous Transmission*. Oxford: Blackwell.

Chapter 3: Respiration

Bartels, H., Bucherl, E., Hertz, C. W., Rodewald, G. & Schwab, M. (1963). *Methods in Pulmonary Physiology*. New York: Hafner.

Campbell, E. J. M., Dickinson, C. J., Dinnick, O. P. & Howell, J. B. L. (1961). *Clin. Sci.* **21**, 309.

Comroe, J. H., Forster, R. E., Dubois, A. B., Briscoe, W. A. & Carlsen, E. (1962). *The Lung*, 2nd ed. Chicago: Year Book Publishers.

Lloyd, B. B. (1958). *J.Physiol.*, **143**, 5*P*.

Chapter 4: Metabolic Rate and Body Temperature

Consolazio, C. F., Johnson, R. E. & Pecora, L. J. (1963). *Physiological Measurements of Metabolic Functions in Man*. New York: McGraw-Hill Co.

Durnin, J. V. G. A. & Passmore, R. (1967). *Energy, Work & Leisure*. London: Heinemann.

Chapter 5: Blood

Barer, R. (1956). *Lecture Notes on the Use of the Microscope*. Oxford: Blackwood.

Britton, C. J. C. (1963). *Disorders of the Blood*, 9th ed. London: Churchill.

Dacey, J. V. & Lewis, S. M. (1963). *Practical Haematology*, 3rd ed. London: Churchill.

Darmady, E. M. & Davenport, S. G. T. (1963). *Haematological Technique*, 3rd ed. London: Churchill.

Chapter 6: Cardiovascular System

DEUCHAR, D. C. (1964). *Clinical Phonocardiography*. London: English Universities Press.

Chapter 7: Sensory Physiology

GREGORY, R. L. (1966). *Eye & Brain*. London: World University Library.

Chapter 8: The Nervous System

BRAZIER, M. A. B. (1960). *The Electrical Activity of the Nervous System*, 2nd ed. London: Pitman.
EYSENCK, H. J. (1953). *Uses and Abuses of Psychology* London: Penguin Books.
EYSENCK, H. J. (1962). *Know your own I.Q.* London: Penguin Books.
HILL, D. & PARR, G. (1963). *Electroencephalography*. London: Macdonald.
WHITFIELD, I. C. (1964). *Manual of Experimental Electrophysiology*. London: Pergamon.

Chapter 9: Alimentary Canal and Kidney

JENKINS, G. N. (1966). *The Physiology of the Mouth*, 3rd ed. Oxford: Blackwell.

Chapter 10: The Effects of Exercise

MOREHOUSE, L. E. & MILLER, A. T. (1963). *Physiology of Exercise*, 4th ed. London: Kimpton.

Chapter 11: Environmental Physiology

EDHOLM, O. G. (1967). *The Biology of Work*. London: World University Library.
LEITHEAD, C. S. & LIND, A. R. (1964). *Heat Stress and Heat Disorders*. London: Cassell.
McCORMICK, E. J. (1964). *Human Factors Engineering*, 2nd ed. New York: McGraw-Hill Co.

APPENDIX C

CONVERSION FACTORS

Pressure	atm	inch water	cm Hg.	Lb/in.^{2}
1 atmosphere	1	406·8	76	14·7
1 in. water	$2{\cdot}46 \times 10^{-3}$	1	0·187	0·0361
1 cm mercury	0·0132	5·353	1	0·193
1 pound/in.^{2}	0·0681	27·68	5·171	1

Energy	BTU	ft.Lb.	J	kcal	kWhr
1 British thermal unit	1	788	1055	0·252	$2{\cdot}93 \times 10^{-4}$
1 foot-pound	$1{\cdot}285 \times 10^{-3}$	1	1·356	$3{\cdot}24 \times 10^{-4}$	$3{\cdot}77 \times 10^{-7}$
1 Joule	$9{\cdot}48 \times 10^{-4}$	0·738	1	$2{\cdot}39 \times 10^{-4}$	$2{\cdot}78 \times 10^{-7}$
1 kilocalorie	3·97	3086	4185	1	$1{\cdot}16 \times 10^{-3}$
1 kilowatt hour	3413	$2{\cdot}655 \times 10^{6}$	$3{\cdot}6 \times 10^{6}$	860·2	1

1 kcal = 427 kg. m

Power	BTU/hr	ft.Lb./sec	h.p.	kcal/sec	W
1 BTU/hour	1	0·2161	$3{\cdot}93 \times 10^{-4}$	7×10^{-5}	0·293
1 Foot-pound per sec	4·628	1	$1{\cdot}818 \times 10^{-3}$	$3{\cdot}24 \times 10^{-4}$	1·356
1 horsepower	2545	550	1	0·178	745·7
1 kcal/sec	$1{\cdot}43 \times 10^{4}$	3087	5·613	1	4186
1 watt	3·413	0·738	$1{\cdot}341 \times 10^{-3}$	$2{\cdot}39 \times 10^{-4}$	1

Length	1 cm = 0·394 in.	1 in. = 2·54 cm
	1 km = 0·62 mile	1 mile = 1·61 km
	1 yd. = 0·9144 m	
Mass	1 kg. = 2·2 lb.	1 lb. = 453·6 g.
	1 g. = 0·035 oz.	1 oz. = 28·4 g.

INDEX